Bear, Believe, Hope, Endure

Stories of God's Grace and Mercy to our Children's Home

Andrew M Lepper

To protect the privacy of those referenced in this book, in most cases names, locations, and details have been changed.

Table Of Contents

Acknowledgements .... *vii*

Introduction .... *ix*

Chapter 1 WWE 1

Chapter 2 We are NOT Orphans 13

Chapter 3 Never Going Hungry 17

Chapter 4 The Pain of Losing 23

Chapter 5 Rescuing 4 Ducks 31

Chapter 6 Duck, Duck, Goose Pond 37

Chapter 7 Kiokee (Ki-O-kee) 41

Chapter 8 Harvesting Bajra 45

Chapter 9 Man 51

Chapter 10 Heston is an Engineer 57

Chapter 11 South Africa Team 61

Chapter 12 John Waller 65

Chapter 13 Prayer Journal 71

Chapter 14 At the Playground 75

Chapter 15 Goat Births 79

Chapter 16 Kanti 83

Chapter 17 Jesus Toffee 89
Chapter 18 Badminton 91
Chapter 19 Baptismal Pool 95
Chapter 20 Boundary Wall 99
Chapter 21 We Grow Grass 103
Chapter 22 S'Mores 107
Chapter 23 Rainstorm 111
Chapter 24 Hot Night in the City 115
Chapter 25 Weather 119
Chapter 26 Blessing and a Curse 121
Chapter 27 Dinosaur Uncle 125
Chapter 28 Extreme Flowers 129
Chapter 29 The Weight of Struggles 131
Chapter 30 Arjun 135
Chapter 31 What are the Small Things Worth? 139
Chapter 32 The Goat and the Rope 143
Chapter 33 Filling up the Love Tank 147
Chapter 34 Welcome to the Family 151
Chapter 35 What does 2017 Hold? 155
Extra Acknowledgements 159
Contact Information 163

This book is dedicated to all of our family that help make our home a reality. Thank you to everyone who donates, prays, and visits. We could not do this without you, not that we would want to. You are a part of who we are. Welcome to the family.

Acknowledgements

I want to personal thank a few people who continually reach out to me to make sure that not only is the home ok, but that I am doing ok. It means the world to me. Special thanks to Robert Dawson, Jay Hutchins, Kelly Parkison, my #LOCO Mastermind, Patty Helms, Ronei Harden, Patrick Henry, Leslie Bumgarner, and so many more. Thank you for your prayers and thoughts.

Thank you to my brother, Scott Cuzzo for his amazing design skills and his willingness to always help out a brother in need.

The biggest acknowledgement goes to Dr. Jason Johnson and his family. There would not be a No Longer Orphans if Jason did not selflessly invest numerous hours in keeping things moving. He is not only the chairman of our Board of Trustees, but he is a non paid COO. He handles every detail in minutia from the American side. We have it easy in India. We get to love on our boys and care for them here. But Jason is the true hero. His sacrificial love towards us is something I could never repay and something I will never forget. He has become my best friend and my brother. I say with no reservation, this organization exists because of the dedication of Jason Johnson. I love you my friend. Thank you for loving us the way you do.

Introduction

Love bears all things, believes all things,
hopes all things, endures all things.
1 Corinthians 13:7

Our home functions on love; more so than financial support. We are called to love one another, and we long for the love from our family in the United States. Love is the standard to which we are called. The title of this book is Bear, Believe, Hope, Endure because that is what love points us to. We are able to bear and endure all the trials that come our way because of love. But we are also able to believe all things and hope for all things because of this same love. Things are not always easy here at the Shiloh Children's Home. In fact most of the time we are enduring some trial. But we can do it because of love. because of Love. It is not trite to remember that God is Love. He is our example and our flag bearer. We are able to handle things that come our way, not because we are innately strong, but because our strength comes from Him; from His love. We want to be an encouragement to you as you read these real life stories and events that have happened to us in the last few months. There are so many more that I was unable to include because of time, brevity, and in some situations, our safety. May these stories strengthen you and remind you that with God, you are able to bear, believe, hope, and endure all trials that you face.

WWE

Anyone who knows me knows that I am the biggest wrestling fan. I remember as a small child my grandmother watching Wahoo McDaniel and Chief Jay Strongbow. I was a huge fan of the NWA. I loved the original Four Horseman, Dusty Rhodes, and Sting. But my favorite wrestler of all time was Ric Flair. WOOOOOOO! I remember watching the first Wrestlemania and everyone since then. But I wasn't the biggest WWF fan growing up. I loved NWA and USWA, but my favorite was WCW (previously NWA). For me, this was where the big boys played. Of course, I still tuned into WWF because more than anything I was a wrestling fan. In the 90's WWF changed names to WWE and they bought out WCW. Wrestling has always been fluid. Wrestlers are independent contractors, so they jumped from company to company when their contracts expired. So it wasn't hard for me to start watching WWE once WCW ceased to exist.

You would think that being in India would make it more difficult to watch and follow wrestling. But it is quite the contrary. This country loves wrestling. Multiple sports channels cover it. Wrestling is shown multiple times a day every day of the week. The Pay Per Views are shown for free. The only downside is that if you want to watch it real time, you have to get up very early because of the time change. I am glad I have DVR because I am not getting up at 5:30 AM to watch wrestling.

On one of the wrestling programs, they advertised that WWE would be coming to Delhi in January. It was the first time in 14 years that they would be here. In the US I can see live events as much as I want. And I did. Within 4 hours of my house, there would be some live wrestling at least once a month. My buddy Willy Lopez and I have been to a ton of live events. Somewhere along the way, we added Kenny, and we all go together now. But I have never been to an event in India and thought it would be amazing.

As soon as the tickets were available I bought 3. One was for me, one was for my brother in law

Tijoy, and one was for my best friend JP. Soon after, I discovered that the dates would not work for Tijoy because he is a nurse in Mumbai and couldn't make the trip just for the event. So I decided that I would make the WWE trip available for one boy to earn. JP said he wasn't too interested, so I decided to make the third ticket available, to a staff member. I didn't know how to make it equitable. How do you choose one boy out of 45 for such a big treat? I wanted to take them all, but at this point, the tickets were sold out for the night I was going. I decided that we would have a Bible memorization challenge. The winner would get the ticket. The same was true for the staff ticket.

I gathered the boys together and told them about the trip. I told them that the winners of the ticket would have to memorize a certain passage from the Bible. Most kids were not interested, but six boys and two staff decided to take the challenge.

For the memorization, I told the boys to pick either ONE of James chapter 1 or Matthew chapter 6. I told them to memorize as much as they could. I didn't want to pick a set amount of verses. How could I pick a boy if all of them memorized the same amount? I told them just to memorize as much as they could, and the boy that memorized the most would get to go.

Boy, did they take me seriously. They were studying and memorizing every available moment they had. I gave them just five days for the challenge. On the day it was time to pick a winner we gathered all of the boys together, including the nonparticipants. I wanted all of the boys to see the hard work that the challengers had put in. I told the rest of the boys that they were going to be the judge. I gave them all clipboards and told them to grade the participants on a scale of 1-10 based on how much they had memorized and how well they did. Since the boys only had Hindi Bibles, I had Susan and JP there to monitor their speech.

I called the six boys and two staff forward. I told them that the competition was about to begin. I wanted to know who chose which book, so I asked the boys who decided to memorize parts of James 1. All 8 raised their hand. This confused me. Why did they ALL choose to learn James and no one wanted to memorize Matthew. So I asked them. I said, "No one wanted to memorize Matthew 6?" One boy spoke up, "Papa, we memorized that also." Let me get this straight- every boy was so interested in the competition, and they all wanted to win so badly, that they memorized everything possible. They had all memorized 61 verses in the span of 5 short days. They still attended school and did all their chores. And they still found the time to memorize 61 verses. It was truly amazing.

I already had a lump in my throat thinking that I had only two available tickets for eight people who had worked so hard. We sat and listened to all 8 recite the 61 verses. This took ALOT longer than we were expecting. We had planned for it to be just a few minutes but it ended up taking almost 3 hours. Some of the boys needed some casual help when they stumbled. They needed us to feed the first word of a verse here and there. One boy stood there confident and recited the 61 verses word for word and did so in blazing fashion. He repeated the whole thing in less time than it took most boys to read 15 of the verses. As they would recite, the rest of the boys were giving scores and taking notes.

When all was said and done, I had eight people who had worked their tails off, but I only had tickets for 2 of them. I felt bad. I didn't know what to do. I didn't sleep well that night wondering how I could fix the situation. Could some of the boys that didn't win at least ride with us and stay on the bus when we went inside the arena? That didn't seem fair.

The following morning Susan and Micah headed off to the Delhi airport to fly to South India. Until that point, Micah had not met his Indian grandmother. I went along for the ride to keep them company. Remember that the tickets for the event were sold out. I had a dilemma on my hands. I had told the boys that I would announce the winner when I returned. Susan and I talked about the tough choice. We even discussed not going at all because it just didn't seem fair. She suggested that I look up the tickets one more time online. SO I did. It seemed that there was one remote location in Delhi that had seven tickets left. That is exactly what I needed.

8 people had taken the challenge plus me made it 9. I already had three tickets, so I needed

six more. I dropped Susan and Micah off at the airport and hightailed it to the ticket venue. On the website, it was listed as CCD ticket sales. I made my way to where I thought it should be but couldn't find the location. All I could see was a coffee shop. I asked around, and no one knew what I was talking about. About that time a worker came out of the coffee shop and asked if I wanted tickets to the WWE event. He wasn't a worker at the coffee shop; he was the worker for the ticket venue. Whew. I had found him. I asked him why it said CCD Ticket Sales on the website. He said that it stood for Cafe Coffee Day, the coffee shop that I was at. His company had a partnership with the coffee company for a remote location.

I thought he would escort me to a side office to do the transaction. Instead, he guided me to a table, in a corner where he had his laptop set up. There was someone already sitting at the table purchasing tickets. I sat close by and waited for the previous customer to finish. I was so tense because the website said there were only seven tickets left and I needed 6. How many would the guy in front of me need? When the time came, he told the ticket worker that guy he needed two tickets. My heart sunk. The worker wanted to confirm that he needed two tickets, so he asked the customer again. The customer surprisingly said that he needed to confirm first with his friend. I was on pins and needles. The customer called his friend but quickly hung up. He said, "My friend cannot make it for that night, so can I just buy one ticket?" I felt like giving him a big kiss right on the cheek. Before long his transaction was done, and I sat across from the ticket worker. "How many tickets do you need?" I confidently replied, "I want all that you have!" His response was, "Wow, five tickets? That is great." "5 tickets?!?!?! That is all you have?" He told me that he would look again. He squinted his eyes and then took out his reading glasses. As he put them on, he said, "Sorry. You want six tickets?" I about had a heart attack right there in that Cafe Coffee Day. "Yes," I said. "But hurry. We can talk about other stuff later. Just sell me the six tickets before it is too late." And he did. And with that, I had the six tickets I needed to make the trip complete. I quickly made my way back home. The 3 hours seemed like 30 minutes because I was so excited to hand out the tickets.

Remember that the boys didn't know where I went. They thought I went to the airport. In their mind, I was announcing two winners, a boy, and a staff member. Even the staff didn't know.

I gathered the boys together and tried to make it as dramatic as possible. You would have thought that I was ready the paternity results on an episode of Maury Povich. I started with the lowest scores. "Ricky, with a score of 7.5 you did an outstanding job. But you will NOT be going to WWE!" Even though one by one the boys faces filled with dejection, I was having too much fun knowing that they would all be redeemed in the end. One by one I got down to the final two boys. They thought that only one would win. "Zachary, with a score of 9.5,

YOU are the winner. YOU will be going with me to watch wrestling!!!!" All of the boys were so excited for him. Of course, they all wanted to go, but they all genuinely care for one another and were happy that he was the winner. I told them that I would announce the staff behind closed doors. The two staff members were there and were confused as to why I didn't do it then. I then turned and walked out of the prayer room, leaving all the boys a little shocked. I stopped just in the hallway where no one could see me, but close enough to hear them. There was positive but confused chatter. "I thought my score would be higher." "I wish I had a second chance to do better." I let them simmer for a moment. Not because I was trying to rub it in, but mainly because I wanted them to savor it more when I made the bigger reveal.

I walked back into the room and told them I had an announcement to make. I told them that I had looked at the score sheets and I had made a mistake. "The winner is not the only one who is going. Everyone is going! Including both staff members." You would have thought they won the lottery. They were so excited. They were screaming and jumping around. They kept saying over and over, "Thank you, Papa!" I was just blessed by all of their hard work and dedication. I will gladly spend every penny I have for them to work like that.

A few days later the 9 of us loaded into a small van. It was pretty packed, but we were elated. We sang songs all the way to Delhi. The driver wasn't as friendly and was noncompliant for most of it. He wasn't too happy that we were so lively. I begged him a few times to stop somewhere to eat, but he was more interested in just getting us to the venue. I was too blessed to get stressed, so I didn't want to make a big deal about it. I knew that there would be something to eat at the arena.

When we arrived at the arena, we took about 200 photos. There was a long walk to get to the front doors after they let you into the main gates. But there were life-size posters of all the wrestlers, so we took turns getting our picture taken with them. We made it to the front, and there was a Dominoes tent at the front entrance. They told us that we could buy pizza but that we couldn't carry it inside. We didn't mind. All 9 of us got personal pan pizzas and Cokes. We all talked about what we might see inside. I was the only one with any event knowledge, so it was both funny and exciting to hear what the boys predicted. One boy thought that a wrestler would personally escort us to our seats. One boy thought that I knew all the wrestlers and I could take them backstage for a meet and greet. I had contacted WWE many times via email and the internet begging them to let our boys come backstage. But I never got a reply. It is ok. The boys didn't know and still had an excellent time. One boy said that when he went to the bathroom that he thought a wrestler might be in the stall and he waited for him to come out. They had creative ideas about what to expect.

We made it to our seats, and the show began. I am not going to give a full detailed report because that is not what the trip was about. The lighting was horrible; the sound was worse. The whole atmosphere was nothing like what I am used to in the US. But that didn't matter. I was enjoying something special with my boys. They had worked hard to earn it, and they were having the time of their lives. We were unable to buy souvenirs because they hadn't expected a big turnout and had not properly prepared. They had sold out before the show was over. All the boys slept like Mac Trucks on the way home, pooped after such a fantastic time.

After we had made it back home, I reflected on the whole experience. It is true that we had the time of our lives. But more important than that was what it took the boys to get there. They had memorized 61 verses. The Bible says the Word of God does not go out void. I believe that at some point in each of their lives they will be in a time of crisis. There will come a day where their back is against the wall, and they do not know where to turn. When they turn inward, they will remember the verses they memorized. They will gain strength and courage from God's

Word that was planted in their hearts so long ago. And that is what this was about for me. Because one day they will not remember much about the trip. They will not remember the wrestlers or the others that went with them. But they will not forget the verses tattooed on their hearts. For me, I would pay everything to make this happen. A ticket to a wrestling event is nothing compared to God's Word that has pierced their minds. To me its worth it.

One final part of the story I would like to mention. A couple of months after wrestling we had another memorization challenge. This time we were taking the boys to an arcade with bowling and cricket batting cages. We told the boys that everyone who met the challenge would be able to go. We set the bar a little higher but still had 15 boys who qualified. We broke the winners up into two trips because we couldn't afford a bus for all of us. I noticed on the first team that only one boy had qualified for both the WWE trip and this one. It was Zachary. He was the one who had memorized all of the 61 verses from the first completion word for word and had done it in fast time. I asked him to tell the other boys how much fun we had on the first trip. He is not a shy kid, but his response was that he didn't remember. I asked him to at least tell them about his favorite wrestler from the first trip. He said he couldn't remember. I asked him to at least say the name of one of the wrestlers that we saw. He said he couldn't remember. I asked him to tell them about what we ate before we went in to see the wrestling. He told the other boys that we ate burgers. I reminded him that we didn't eat burgers. We ate pizza. "Oh yeah," I asked him if he could remember anything about that trip and he just looked at me blankly. I was so confused. Had the trip been nothing that I remembered? After 2 minutes I asked, "Well, do you at least remember part of the bible verses?" "Oh yes Papa, I remember it ALL!" And with that, he spent the next few minutes quoting ALL 61 verses verbatim. This was what it was all about. We still do these memorization challenges. We still have fun incentives. Why not? But for me, it is not about where we go or what we do. It is about encouraging the boys to store God's Word in their heart. And for that, I will pay anything in the world.

We are NOT Orphans

I will not mention names for this story because I do not want to bring unnecessary attention to any team that has visited us. It was just a simple interaction, but it melted my heart.

We had a team visiting us. They had been here for a few days and were still with us when Sunday rolled around. So naturally, they joined us in our morning service. We proceeded with our regular routine, and then towards the end, we asked the team if they wanted to share a little bit about their trip. One by one they came to the front and told the boys how much they enjoyed spending time with them. It was an excellent time.

Towards the end, someone came to the front and through translation told the boys that it was a blessing to be able to visit our orphanage. There was an awkward silence, and then the boys started gently laughing. It made the team member uncomfortable, so they repeated their last sentence, "Thank you for letting us stay at your orphanage." The boys began to laugh even harder. The team member was visibly embarrassed because they were unsure of what they had said that was so funny. I was sitting in the front next to the person, so they looked at me and asked what they said that was so funny. I looked at the boys and locked eyes with Raphael. I told the team member to ask him what was so funny. They asked what they had said wrong. Raphael's response sent chills down my spine.

Honestly, I didn't know why the boys were laughing. I thought maybe something was lost in translation. But our boys are kind and respectful, so I know they didn't intend to shame anyone. Raphael stood up and said, "You told us thank you for letting you stay here at our orphanage. This is not an orphanage. Orphan has no father. We all have a father. He is sitting next to you. This is a children's home. This is Shiloh Children's Home. We are the Shiloh boys, and this is our home. We are trying not to laugh, but it is funny that you are telling us in front of our father that we stay in an orphanage." It was like Raphael did a mic drop. I know my mouth dropped. I was stunned.

That was one of the sweetest interactions I had ever been a part of. It still gives me chills thinking about it, or as the boys say, "goose pimples." It finally dawned on me that the boys "got it." They knew that we were a family. They knew that we were here for them and that we weren't leaving.

I have described many times to people that we have an orphanage, despite the fact that our name is No Longer Orphans. The sole reason is that it is just easier for people to understand what an orphanage is. But no matter what we call it, or what name it bears, to my boys, this is just HOME. It is not an orphanage or even a "children's" home. It is simply HOME.

Never Going Hungry

This is one of those hard to believe stories. But trust me that it is entirely true. It feels like it is borrowed from the pages of George Meuller's autobiography, but I attest that it happened to us.

In 2015 there were 54 times where at some point during the day we wouldn't have enough food to eat. You didn't misread that. There wasn't a typo. 54 times.

If you do the math, that is right at once per week. So on an average of once per week for a whole year, we were put in a situation of going without food. But this is not from neglect or lack of planning. In fact, we try extra hard for this never to happen. That is why we grow wheat, corn, mustard, and other crops. It is why we have goats, chickens, ducks, geese, and water buffaloes. There were times where our city was on strike, and although we had money for food, there was no way to get it. There were times of monsoon where the markets were flooded. There were so many ways in which we had made plans to fill our pantry, but it was still barren.

54 times without enough food to survive. But can you guess how many meals our children missed? The answer is zero. Even though there were 54 times that we were without food, there was not a single time that we missed a meal. But how?

Let me start by saying that I believe these small acts of miracles were done for a particular time to recalibrate us towards a God-centered focus. I truly believe that God did it "this way" to bring Himself more glory. So any preparation on our part was futile anyway. This was apparent when we would buy extra monthly rations, but they would still run out. It was like the reverse of the story of Elijah and the widow in 1 Kings chapter 17. In that story, God replenished her flour and oil every night. She never ran out as God had promised. For us, it was the opposite. We would close and lock our store room confident that we had enough food for the following day, only to open again and realize the food would run out long before we filled all our hungry bellies. It felt like food would vanish in the night. But that was highly unlikely. We always locked the storeroom, and I was the only one who had the key. And every time we discovered that we were lacking, it was soon followed by small acts of kindness.

We have had times where people would show up without warning to provide breakfast for the children. We have had local shops call us to tell us they had a donation of food for us. We even have a local vegetable vendor who would randomly give us all the leftover vegetables at the end of the day. But we never knew when it was coming.

Don't forget our animals. We have chickens mainly for the eggs and the water buffaloes for their milk. There were so many days that the chickens just simply didn't give any eggs, or gave a minuscule amount, only to be inundated the following day with more eggs than we needed. The same was true for the water buffalo and the milk. I can say that there were days where it just didn't make sense that there was no food available. But in every situation, God provided in miraculous ways, and our boys never went hungry. It became a challenge to figure out how God would provide for us and we welcomed it, knowing that God would meet our needs.

I want to share with you one story in particular. Bear in mind that not every child knew that we had these struggles with food. Some of our boys come from poor backgrounds, and the last thing I wanted was for the smaller boys to be food insecure, wondering where the next meal would come from. That is our responsibility, and we didn't want them to worry. But we have made some of the times aware to the older boys so they could pray accordingly. One morning I woke up to discover that our pantry was bare and we had nothing for breakfast other than spices. I called 5 of the older boys aside and told them the situation. I asked them if they would join me in prayer and they agreed. We sat down and began to pray. I was just getting warmed up when they got up and started to leave. I asked them why they were leaving so soon, and one boy responded, "Don't worry Papa. God will provide. We know he will. We believe He will so we are going to get ready for school." I wasn't as convinced quite yet, so I decided to pray as I walked around the building. The boys would soon be leaving for school, so I went to our front gate and unlocked the thick chain.

No sooner had I walked back to our front door did a small car come screaming through the front entrance. The way the car came to a stop, I was sure that the driver used to be a stuntman on The Dukes of Hazzard. He got out and quickly ran over to me. I was unsure of his intentions, so I positioned myself between him and our front door. He immediately said, "Sir, you have to help me." I was so confused but curious about what this man wanted at such an early hour. It was a little after 6 am, and the sun had risen not long before. "Please help me sir or my wife will kill me," I asked him what in the world he meant. He told me that he had gone to the vegetable market bright and early to get his families fresh food for the day. He had also picked up a bag of rice. As he was driving home, he noticed that there were two bags of rice in the back. He told me that his wife would never believe that he didn't buy it and that someone accidentally put it back there when they were loading up the other bag. He told me that from experience he knew that his wife would fight with him and he just didn't want to deal with it. So he asked me if I would please take it. I told him that I would gladly take it because we were not sure where our daily food would come. He gladly handed it over, and he was on his way.

But before he got in the car I had one question to ask. I said, "What made you stop here? I have not seen you before. Why did you come to us?" He told me that he had stopped along the side of the road and got out to put the rice on the side. He knew that someone would come along and take it and he wouldn't have to explain anything to his wife. But then he said, "As I was about to take the rice out, I noticed the sun shining through the plus sign on the top of your building. Something inside told me that the "plus sign people" would take my rice. So I came to you, and you took it." Once again God had provided in a miraculous way for us. He used a nonchristian passerby to provide our daily food. The man had noticed the cross on our building, and God used him to provide for us. And by the way, it wasn't just a small bag of rice. It was a 50-kilogram bag of rice which is 110 pounds. SO not only did he provide our daily meal, but we ate for days off of that one act of kindness.

Through 2015 we had crazy ups and downs. But God showed up in miraculous ways, each time letting us know that we were loved and that he was taking care of us. We have not had any days without food in many months, but God has cemented in our hearts and minds how far He will go to meet our daily needs.

The Pain of Losing

In my first book, Birds, Buffaloes, and Birthday Bread, I had a chapter entitled "Addition by Subtraction: Why I Hate Math." It told of us suddenly losing three boys only to gain three more in a matter of a few weeks. It is still hard when I imagine the three boys that left us. I still remember their faces.

That was easy compared to what we faced in 2016. In the span of a month, we lost 13 boys. THIRTEEN. They all left for various reasons. Similar to the last time, some went home for the summer and just never returned. A few returned but no longer wanted to be here. We love these boys with everything we have. But part of that love is trying to instill discipline in them that they will not get elsewhere. We have expectations of every boy. First, a home our size without expectations or discipline would collapse on itself. If we had 40 boys running around with no guidelines, it wouldn't be very pretty. But most importantly, we have disciplines and expectations because we believe our boys can make something of themselves.

Almost all of the boys come from destitute labor families. We have the crazy belief that they can rise above that and be anything they set their mind to. They can be doctors, lawyers, and engineers. They can be artists, musicians, teachers. There is no limit to their dreams. But we put expectations on them because otherwise, dreams would never become a reality.

Having a dream in and of itself leads to nothing. You must act on the dreams and put in the sweat equity to bring dreams into reality. I wish I had a dollar for every time I have talked to the boys about putting the work in for their dreams. Many times boys have told me that they want to be doctors or engineers. I sit them down and pull out their grades. I ask them how they could get accepted into medical school if they fail math and science. But it doesn't just end with my questioning their studies.

I ask them what I can do to help them get

better. We just don't tell them what it takes to excel in school. We provide the means to make it happen. We give them adequate study time and hire tutors to assist them. I am foolish enough to think that even though you may not be naturally gifted, you can work hard enough to make something happen. Even if that means working harder at math and science so you can become a doctor.

We believe in our boys, and our expectations show it. But some boys have never had anyone push them to greater things, so they feel uncomfortable. They gravitate towards doing as little as possible, but we expect them to do more. With that said, there is a delicate balance between pushing the children enough, and overdoing it and pushing them too hard.

Kids deserve to be kids. They need time to laugh, play, and explore. We afford them this. This is just part of growing up. But study time is study time, and we expect them to take it seriously.

Because we have so many boys, we have to keep a schedule. We cannot have 50 people living under one roof keeping 50 different schedules. So we have a time for waking up, prayer, meals, play time, and study time. It becomes chaotic very fast if we didn't. Can you imagine a 3rd-grade class where the teacher let the students do what they wanted when they wanted all day long? What would get done? The same principle applies to our schedule.

The boy's families have entrusted us to care

for their boys and to provide them with an excellent education. And we rise to meet that challenge. But the sad truth is that even though their family expects us to provide this for them, they do not expect much from the boys themselves. When the boys return home, they run amuck. There is no schedule. The boys come and go freely and wander the streets. Their family doesn't feed them like we do here and most do not care where they are through the day.

For some boys, when they were home for the summer break, they had zero structure or discipline. So when they returned to us, they didn't adjust well. We tried to have as much grace with them as possible, but it didn't matter.

One boy, in particular, began to run away frequently. I cannot disclose full details, but it became almost an everyday thing. He would run through the fields and jump the fence first thing in the morning when the other boys were getting ready. We would only notice once the boys were loading up on the bus to go to school. We would search around and quickly find him. He told us he just wanted to go home. He had no mother and father, and his grandmother abused him, so we asked him why he wanted to go back so bad. He told us it was because his grandmother lets him do anything he wanted. We asked him about meals. He said she didn't feed him, but he would rather go hungry without rules and than to eat healthy food but have to study all the time. So he just kept running away.

This is hard to write. Partly because we fear of what people may think of us. But mostly because we are left with a huge void. We keep asking ourselves if we did enough; if we could have done things differently. But we talked at length with all the other boys, and they had different feelings. All of them told us that they loved the discipline because they have never had that before. They said that they love it because it shows them that we care about them. They take pride in their school, their daily duties, and their relationship with us.

The remaining boys were a comfort to us as we dealt with the loss of 13 boys in a month. As mentioned before, some just decided to stay home instead of returning from summer break. A couple ran away, and we had no choice but to force their immediate family to take them back. I cannot go into detail, but they put us in a position of jeopardizing the whole children's home, so we had no choice but to send them back.

There is one story in particular that broke my heart. There was one boy who kept running away. We had been told by the police to get the situation fixed, and fast. We asked the boy's father to take him home, but the boy's father refused. We then asked the boys father to sign an affidavit that said that if his son ran away, that we couldn't be held liable (the police demanded this.) The father agreed. But I was not expecting what happened when his father arrived.

His father barely made it through the door

before his son ran to his arms. But the father would not embrace him. Instead, he pushed him away. Instead of comforting him, he began to shout at his son. His words were like arrows, piercing the ears and hearts of his nine- year-old son. "Stop it! I didn't come here to take you home! I do not want you! You are not welcome in my home anymore. I have a new family. You need to stay here with this new Papa and Mommy who love you more than me! Stay with them because I do not want you." And with those last words, the father threw the affidavit in my direction and walked out, never to return.

We still have not seen him. His son was still standing in the doorway, shocked and sobbing uncontrollably. His two older brothers were standing behind me just as shocked. Until this point, they had not discouraged their brother from running away. Maybe it was because they wanted a little more freedom if their brother left. Maybe it was because they felt their younger brother would be better off at home. Who knows? But they were indifferent up until that very moment.

As their brother crumbled onto the floor in a sobbing mess, I knelt beside him. I decided not to say anything. I just simply put my arm around him and embraced him. I was as shocked as he was and that is the only emotion I could muster because I was choking back my tears. We knelt for what seemed like hours. After some time his brothers came and joined us. I didn't have to say anything. They said it all. They said, "This is our home. This Papa loves us more than anything. Let us be happy with our home here. Let us forget our old life. Mother left, and Father treated us badly. Let us take the love of Papa and Mommy."

And at that moment, this young boy became strengthened. He became resolute. He hugged me like there was no tomorrow. And he moved forward. We have not had a single issue with him since that day. His grades drastically improved. He has become a leader of the younger boys. You can tell that he truly feels like he belongs; like he is home. While fully correct, it's not the same story for all the boys. A few have left for various reasons. And we feel the loss. To go from 45 boys to 32 in less than a month leaves a void; both physically and emotionally. We choose to continue to fight. We must fight, for the boys that remain.

Not a day goes by where I do not think of the boys that have moved on. I question what we could have done differently. I may never know. I will not lie and say that we have moved on from this. How can we? This affects us. This affects our relationship with the boys who are still here. How could it not? I find myself wanting to withhold emotions and love because I am afraid of getting hurt again. This is just the honest truth. But I must remember that one day all of the boys will move on. Even in a perfect world, they will go to college; get married; start a family. Would I knowingly withhold the only love the boys receive because I am afraid of being hurt? This past year was the hardest of my life. Mostly because of the loss of boys. But I have rededicated my life to loving freely with no strings attached, even when the love is not returned.

Rescuing 4 Ducks

Recently Susan and I were in Delhi visiting friends. We had a great time hanging out with their family. Susan is from Kerala and so are these friends. They took us to a great restaurant to eat. Micah discovered pickled onions, and that is all he would eat. They took us to a small Kerala shop nearby, and Susan bought some spices and other packaged goods from her home state. South Indian food is a little different than North Indian food, and the recipes and ingredients are a little different.

They told us about a local open air market that had Kerala vendors. They sold specialized fruits and vegetables as well as more fresh spices. It was a super hot day in Delhi- easily 110 degrees.

Micah had had a full day up to that point and was sleeping it off in the back seat, so I decided to stay with him instead of waking him up. The A/C was struggling to keep up with the humid, high temperature. It was muttering and sputtering, but it was still cooler than being outside.

Susan promised to take only 10 minutes, but I knew better. Around the 30 minute mark, the air conditioner stopped blowing cold air altogether, and Micah quickly woke up. He was hungry, so I texted Susan to return immediately. I was not prepared for when she came.

She was holding a dirty, skinny duck. Duck is a common luxury in Kerala. Many people eat it as a fancy alternative. Kerala is known for its rivers and waterways. Along many of these rivers, you will find pop up duck operations. They set up nets measuring 20 feet by 20 feet within the water and above it. The nets aren't meant to catch wild ducks. It is intended to create a temporary pool for the ducks to swim in. I have seen as many as 500 ducks swimming around in a single one of these duck net ponds. You can pull up and buy a duck. The operator has a tiny wooden shack beside the road where he can butcher the duck on the spot. Now that is some fresh tasting duck! Duck curry truly is amazing.

But here in this market in Delhi, there was no duck pond. There were just ducks that were too tightly packed into small cages. Four ducks were packed into an enclosure that even one shouldn't have been in.

Don't get me wrong- Susan and I are by no means vegetarian. But we value animals. We have a farm so we can properly care for the animals we own and eat. We treat our goats and chickens great, not just because it makes them taste better, but because it's just the right and humane thing to do. We even cook about 50 pounds of food for our water buffalos every afternoon on an open fire. We love our animals.

So when Susan saw the ducks, her immediate desire was to rescue them from the cage they were in. What a horrible life to be stuck in a cage too small, waiting to die. This is one of the main reasons we began to raise our chickens. I could no longer buy lethargic chickens that lived their short lives in small cages. Our chickens live healthy, long lives. In fact, we do not butcher them until they are beyond their egg laying capacity. And so it is with the ducks.

Susan's first inclination was to fatten the ducks up for a couple of weeks and then butcher them. We love duck curry. But that soon changed. When we made it home from Delhi, our rental car just smelled rank from these four dirty ducks. We put 2 square buckets of water in one of our guest bathrooms, and within minutes these ducks had bathed themselves clean. They were frolicking in the water (probably for the first time) and doing duck things. The water went from crystal clear to dirty black in less than 5 minutes. We decided to keep them in the bathroom overnight so they could acclimate to their new life with the chickens during daylight. Our roosters are both friendly and territorial, so we didn't want them to freak out when four random ducks showed up in the middle of the night.

We discovered that we had 3 Daisy's and 1 Donald Duck. They did well the following day with the chickens, but we brought them back inside for the next three nights for their safety.

On the fourth afternoon, we looked for the ducks in the chicken pen but couldn't find them. We feared they had either been stolen or flown away, or been taken by a predator. We looked EVERYWHERE. Well, everywhere except the chicken house. We soon discovered after looking everywhere else that they had become so accustomed to the chickens that they just went inside for the night when the chickens did. It was like an Old MacDonald's Farm version of where's Waldo. It was near impossible picking out four ducks in a sea of 500 chickens in the chicken house.

The ducks soon began to thrive. We were very quickly faced with a tough decision- to eat or not to eat. It didn't take much for us to decide NOT to eat. The Fab 4 had quickly become more than our pets; they had become like our mascots. Every morning, soon after we opened up the chicken house for the chickens to peck around their yard, we would let the four ducks would waddle out of the fence to march to our small water buffalo pond and swim to their heart's content before waddling back inside for the heat of the day. It was a sight to see! These ducks went from being stuck in a cage with no water, being miserable, to being the mascots on our homestead farm. They went from the possibility of being dinner to being a part of our family. If you visit us, you will surely see our four miracle ducks waddling around like they own the place.

Duck, Duck, Goose Pond

Once the ducks were acclimated, we began to think about how to give them fresh water so they could enjoy themselves. I mean, they are ducks after all. They weren't intended to perch in a tree. I ordered a couple of vinyl kiddie pools from Amazon India. These worked surprisingly well. With only four ducks, they had enough room to swim a little and bath. It was still better than them having to share a small plastic tub.

Not too long after we got the ducks, I got bit once again by the feather bug. By that, I mean that I had the desire to get more feathered friends. Up to this point, we had taken in over 1000 chickens and now four ducks. I thought the next logical step would be to get geese.

I have a buddy in the U.S named Justing Rhodes. He has been dubbed the "Permaculture Chicken Ninja" and also the "Birdman." He has a goose. But he doesn't have it just for fun. His goose has a purpose. His goose is a "guardian goose." He houses his goose with his chickens to cut down on predator attacks. Geese are loud, and so they are perfect alarm systems in case something is trying to attack the chickens. I decided that it would be perfect to get a guardian goose.

Since day one our chickens have been regularly attacked by wild street dogs, with nothing to deter them. One day in 2015 we saw the destruction that one dog could do. We had just taken in 300-month-old chickens. We had them fenced off right next to our back door so that we could keep an eye out for predators and thieves. We heard a horrible racket going on and quickly ran out. It didn't take a minute between the time we heard the sound until we were out there. But it was too late. One crazed lone dog had killed 30 of our chickens in a short period of time. Maybe the chickens had been making noise, and we hadn't heard. Regardless, we had lost 30 chickens senselessly even though they were right next to our back door.

I decided to see if geese may be the answer we were looking for. Susan heard about a man in a nearby village selling a goose, so she went and checked it out. She came back with two beautiful geese, a male, and a female. They fit in immediately and let out some unusual honking sounds as I am sure you could imagine. I knew they would be an excellent alarm system for our chickens.

And it worked. In the six months that we have had them, we have not lost a single chicken to a predator. They are the best guard dogs I have ever had. And that includes the St Bernard!

The geese were, of course, bigger than the ducks and had a greater need for more water. Plus, within hours, their toenails had pierced the hard vinyl bottom of the kiddie pool, and it leaked out everywhere. Something else had to be done.

I decided to mark off a section of the chicken pen and dig a pond by hand. It wouldn't be too wide, and we would only dig it four feet deep. I spent the better part of a morning getting everything played out.

Our tools in India are old school, so I didn't have a standard shovel. It took a couple of hours just to remove the top layer of grass to show the surface dimensions I wanted. The boys were off at school, and in the heat of the day, I decided to take a break and get some office work done.

Shortly after the boys came home from school, I heard a lot of commotion from outside my window. I went to check it out. The boys had been so interested in the project that I started, that they wanted to be a part of the action.

They had rounded up all of the makeshift tools we had. They were on their hands and knees digging the pond. We spent hours trying to dig. We were making progress, but I knew that it would take a long time. The boys were not forced to help me.

They have an amazing work ethic and were helping me even before I knew they were.

You can see videos of my original work as well as their digging on my youtube channel, youtube.com/theandylepper

The next morning as the boys went off to school, I realized that it would take longer to dig this pond than it was probably worth. Susan and I inquired about a JCB backhoe. The driver charges you either by the cubic foot of digging or by the hour. We calculated and realized that it would be super cheap to dig not only this pond but a new pond for the water buffalos. We ended up spending less than $100 total. The duck and geese pond is 30 feet long by 15 feet wide by 6 feet deep. The ducks and geese love it. We have even begun to stock it with local fish.

Kiokee (Ki-O-Kee)

In middle school and high school, I had a friend in my youth group at The Baptist Church of Beaufort named Michelle West (her last name is now Parnell). She was awesome, and we have kept in touch through the years.

In 2015 she asked me if I would be interested in sharing about the children's home to her WMU group. WMU stands for Women's Missionary Union. Michelle is a member of Kiokee Baptist Church. It is Georgia's oldest continuing Baptist church. It has a rich heritage, and I was eager to not only see the old buildings but to also share about our boys.

We had an excellent meeting, and afterward many of us went out to eat. I remember a few of the ladies from that day, but one, in particular, stood out. Her name was Marcia Bailey, and she was the youngest most gorgeous 70-year old I had ever met. At lunch, she matter of factly she stated that she would bring a team to visit us in India within the next six months. I had no reason to doubt her. I had just met her and was already in love with her enthusiasm.

But I have seen more than my fair share of people with good intentions later inform me that the timing didn't work out or they couldn't get enough interest for other people to come. I am not an ageist, but we have never had someone in their 70's even visit us, let alone lead a team her first time to India.

But Marcia is persistent. She followed through, and in April of 2016, she helped lead a team from Kiokee to visit us. Initially, we thought it might be a little earlier in March, and I had already had speaking engagements in the US for the first week of April. So I knew I was not going to be in the country when the team arrived.

But I met the team in Georgia a few days before they left for a team and trip briefing. I was jealous that I would not be in India because I knew this was going to be a fantastic team. The youth pastor, Jonathan Melchior was helping lead the team with Marcia, and he was awesome. Two other people stood out to me; Lisa Harvey and Michelle

Tordoff. I just knew that they were go-getters and would impact our boys. Michelle took 3 of her kids with her, Michael, Praise, and Timothy. Lisa took her son Charlie.

We loved the whole team, but we have bonded with these two families. Lasting friendships were made on that trip. Not a day goes by where at least one of our boys does not wear their Kiokee trip shirts they were given.

I made five trips to the US last year, and each time I stayed with Lisa. She has allowed us to use a room in her house not only for our personal belongings but also to do Chunky Junk. Michelle Tordoff volunteers with us weekly in helping No Longer Orphans and Chunky Junk with administration duties. And each of my trips I got to spend quality time with Marcia. She is like a second mother. She is awesome, and I thank God that He brought her and Kiokee our way.

We have grown to love Kiokee and its members greatly. They are a vital part of our family. By the time you read this, they will have made their second trip to serve alongside us. They will have spent a week loving on the boys, taken the time to ride elephants, and journeyed to see the Taj Mahal. And surely, they will have strengthened the deep bond they share with our boys.

When Kiokee visits, it doesn't feel like we are hosting a team. It feels like our family has come for a visit. To us, Kiokee is not just a church. They are not just a partner. Their members have been grafted into our family, and we have been forever changed by the kindest they have shown to us.

Harvesting Bajra

When we first arrived at our farm, we had a sharecropper who partnered with us. He was a seasoned farmer who knew what to plant and when. Our agreement was that we would provide the land, the water, and the electricity for the water pumps. He would provide the seeds, fertilizer, tractors and labor. Come harvest time; we would split the crop 50/50.

At the beginning of a season, he would show up with a rented tractor and plow the fields. We have roughly 10 acres of plantable fields. He would have a team of laborers turn up, and they would plant all the seeds. And that was the bulk of the work for the crop. Depending on what he had planted, he would flood some areas of the field. He would create a channel down the middle of the main field all the way to the back wall of our property. He would fill the channel with water, and it would flow all the way down. After it had flooded that patch of the backfield, he would dam up a section, and the field in front of it would receive water from the channel. He kept doing this until it was all watered. It was laborious, but it was a one man job for someone with the skill and knowledge to do it.

The sharecropping farmer had many fields other than ours, so he had his father come and be the caretaker for our fields. Systematically day by day his father would water the fields. This is all done by hand; back breaking labor.

We do not use pesticides or chemicals, so many "weeds" would grow along with the crop we planted. Couple that with the fact that the bags of seeds had lots of other stuff in it, and it wasn't hard for evasive weeds to pop up amongst "the good stuff."

But the good news is that although we didn't intend to grow these things for ourselves, these "weeds" could still be fed to livestock. The farmer had a key to the lock on our back gate, and he freely passed it around to people who were willing to remove these weeds. Daily there would be women who would come through the back gate and spend all day gathering greens for their water buffalos.

That is all well and good, but many of them didn't discriminate between what we were trying to harvest and what they were allowed to take. So day after day they would just gather everything in site. They would collect the good stuff but cover it with the weeds thinking we didn't notice. I am not sure what agreement they had with the sharecropper, but he turned a blind eye every day when they left. That may not seem to matter, but in the end, it does.

Of course, we want to be able to help all of those around us, but once it came time to divvy up the harvest after we got our 50% of what was left, we would barely break even after accounting for the water and electricity. In India, people do not confront someone that is doing something like that so I knew that we would just have to split from this farmer amicably. He wasn't thrilled, but there is no way we could lose money by farming. Growing crops is meant to be a blessing for us, not a hindrance. We parted ways.

I decided then, instead of doing a 50/50 split that I would just charge a flat rental fee to any farmer who wanted to use the field for one year. It was no surprise to me, but the sharecropper didn't want to do that because he said he wouldn't make enough money. I wanted to charge enough money that would account for the use of our water, land, and electricity while still giving the farmer enough profit once the crops were harvested.

He was truly the one who had the most to lose. If the crops were lost, it wouldn't be our fault, but I didn't want the farmer to lose it all. I found one local man who is in the Army. He was willing to give it a go, and we settle on a fair and amicable price. He would have his brother handle all of the actual farming duties. And so it was.

He had a great year, and his harvest was plentiful. He made a significant profit, and we were also satisfied. But then he got transferred, so I was back to square one.

I had to decide what we would do next with the land. There are not many variations on what is planted in our area for mass crops. People do not stray far from the planting of corn, mustard, guar, and bajra. Bajra was the next crop to be planted so I decided that we would do it ourselves.

The staff agreed, and we were excited to be doing something more sustainable for ourselves. Bajra is better known in the United States as pearl millet. People around here just call it Indian corn. It is a typical food in village areas, and it is also used as feed for livestock. Our chickens love it.

We decided that we would primarily grow it as food for our chickens and water buffalo. We did the same thing that the sharecropper had done. We rented a tractor and plowed the fields. We all had a part in planting the seeds. And then we waited.

We watered it one time. Bajra is a very resilient crop and is known for being able to withstand drought or even flood. We would rather save water than to waste water, so we didn't water it as much as we did other crops.

After four months of steady growth, it was time for harvest. But hand harvesting 8 foot long stalks is tiresome. The harvest is done in stages. We first cut down every stalk and left it there as we made a steady cut through the field. We then went back and cut the head of the bajra off the stalk and made two pile. One pile we stood the stalk straight up and made a pile that looked like an Indian teepee. We laid down a plastic tarp and laid down the heads of bajra on it to dry. We let both the stalks and the heads of bajra dry for two weeks.

The South African team was here when we decided to clear the fields of the stalks. Our field is pretty deep, and we had piles of the stalks everywhere. The South African team and our boys brought all of the stalks to the front of our building to further dry.

Once they dried, we chopped them up, and they became chaff that we can feed the water buffalos. The heads of bajra needed another week. When the time came, we rented a tractor with a harvesting implement to help us separate the grains of bajra from the head. This separation process was a fascinating thing to observe. We would fill buckets with the bajra heads and pass them to the back of the separator. It would send the grains down one shoot and then grind the head into dust and shoot it out the other side. You should check out the youtube video we made of the harvest.

I can honestly say that this was hard. There is a lot of back breaking work to farming, especially when you cannot afford the right equipment. It is our desire to one day get a small tractor of our own. The time we spent working elbow to elbow in the fields was great. We came away with a newfound respect for the food we eat. We take pride in being able to produce things for ourselves. We have many plans for growing our farm, but we will never forget the first time we planted and harvested our own crop, bajra.

Manu

Let me tell you about Manu. He is Susan's nephew, the son of her older sister. His father died when he was very young. His mother is uneducated and works for a family in the Middle East.

Manu bounced around from family member to family member his whole life. His life growing up in Kerala did not have much structure or continuity. There were weeks where he stayed at a different place every night in the village.

He never finished school. He is still very immature. But that is because mostly no one has ever modeled for him what it means to be an adult;

what is expected of a man. There are not many men in Susan's family; not good ones at least. Her father left the family and died homeless a few years ago. Her uncles are out of the picture. Her grandfather passed away many years ago. Her older sister's husband was not a good man and died also. I came along 14 years ago, and I was the only man in the family until her younger sister got married. All that to say- Manu has never had an adequate role model.

A couple of years ago he began to be quite a handful to my mother in law and the other families that he stayed with. He was 18 at the time, and he wore his welcome out as fast as he arrived. He had become disrespectful and had no direction. In essence, he came to us because no one else wanted him.

It is not that we hadn't always wanted him, we did. But he is from Kerala which is about as far away as you can get from us in India. He didn't know Hindi and didn't want to come. But once we realized that we were his last and only chance we quickly begged him to come to us. I had one person ask me, "Why do you want HIM? He will never amount to anything." At that moment I realized that I would do anything to give Manu a better life. Every kid deserves a chance for a better opportunity.

So we welcomed Manu into our home. In one was it was an easier transition because he was already family. But it was equally as tough because he tried pushing our buttons every chance he got. This was partly because he just wasn't raised to show respect to anyone. He just didn't know. Teaching simple things to Manu was like training a chicken to swim.

(I personally do not know how hard that is, but it has to be near impossible, right?)

When they dropped Manu off, some family stayed with us for a few days. On the second day, Manu asked for the keys to our old van so he could wash it. I didn't think anything of it at the time. Within 10 minute my mother and law ran in screaming that Manu was driving and he would kill himself.

I went to see what the fuss was all about. Manu was in our front driveway just driving in big, slow circles. He wasn't causing any harm. I flagged him down and began to talk to him. I asked him who taught him how to drive. He told me that no one had taught him, he had just learned by observing others. I said I was impressed at how well he was driving but scolded him because he didn't take permission to drive around, he only took permission to use the keys to open the door to wash the car.

This quickly became a pattern with Manu. He would do the most fantastic job around the farm, but he did things without asking or even letting anyone else know.

One day I woke up and found that Manu had hand dug a huge, massive hole in the middle of our field. He had spent ALL night digging it. But it was right in the middle of the field that we were preparing to plant. I asked him what on earth he was doing. He said that he noticed that the water buffaloes needed a better pond, so he just decided to dig one- in the middle of our crop field. This is just how Manu's brain worked. He has an amazing amount of self-motivation. The problem is he never clears his endeavors with us before he starts acting on them. What Manu lacks in common sense, he makes up for in initiative. He was not happy that I made him fill in the pond that he had just worked so hard to dig. But I told him that had he asked; I would have shown him the spot that I had already marked off to dig a new pond. He was shocked to learn that I had plans to dig a pond.

I told him that this is how people function- through communication. When we adequately communicate with one another, we are all happy and more can be accomplished.

I wish I could report that things ran smoothly after that, but they didn't, and that is ok. Manu is a work in progress. He had 18 years of poor communication. 18 years of no direction and people not having much belief in him. We desire to give him direction, and we have a great belief in what he will do in this life. But it takes time for him to believe in himself and to know that we believe also.

I admit that I am extra hard on Manu. That is because time is precious and we are trying to get him prepared faster than anyone else. If we are not able to instill some simple life lessons in him soon, we may not have the chance later. These life lessons include time management, discipline, teamwork, maturity, and so much more.

When Manu first came I asked him what he wanted to do with his life. His response was that he wanted to be a driver. I can get behind that. That is a noble profession. Manu didn't have much male mentorship in his life, but it didn't take me long to figure out that most of the men that he looked up to in Kerala were drivers. So I affirmed his choice and set out to figure out how we can help him. Being able to get a license is not the same as being mature enough to drive a car. We are still dealing with some maturity issues with Manu because for his whole life he has never had responsibility and has been treated like a small child. He is of driving age but not mature enough to drive yet, in my opinion.

One of the ways I wanted to teach responsibility and maturity was through our farm. I gave Manu full responsibility of our chickens. He became responsible for feeding, watering, shelter, gathering eggs, and cleaning up their area. He is in charge of selling the eggs and butchering the chickens when someone needs it.

Manu is an extremely hard worker but couple that with his supernatural stubbornness, and we have interesting encounters. I am sure there are days that he thinks I am Satan himself. I try to show grace to him when needed, but I am tough on him because I love him enough to do so.

I will only say, that even including our staff, I am the person with the most faith in Manu. I know he will be a great man one day. There are days when he is the most immature in our building, including the small boys.

But as I write this I think back in the two years he has been with us. I reflect on where he was when he came to us- unwanted, uncared for, no discipline, very immature, and no direction to get where he wanted to go. All he knew is that he wanted to be a driver. He had no clue how to get there, and no one to help him.

But we cannot concentrate on where we come from. Or even where we are. We must focus on who we should become. We are all works in progress. Manu is a work in progress. We worship a God of second chances. He never leaves us in our past sin or dwells on it. God patiently looks to who we can become; who we should become. God's patient love never gives up. That is my desire for Manu.

I know that one day it will be said of Manu- "That man was an orphan, but now he has a successful transportation company (or whatever else he does). He was an orphan, but now he drives for Kings." But my prayer for Manu most importantly is this, "Once he was an unwanted orphan, but now he is a Godly man who loves his family and gives glory to God. he once had no direction but now is a leader for many."

Heston is an Engineer

Heston is one of our original boys. He has been here since the beginning. He is now ten years old. Just when you feel like you know someone well enough, you learn something so new it leaves you speechless.

Shortly before Christmas one of the staff brought Heston to me late at night. I knew from the look on the staff members face, and also Heston's face that something bad must have happened. I immediately asked him if he was ok or if he was hurt. He said he was fine. The staff member told me I would never guess what Heston had done. I was genuinely curious.

Heston's hands up until that point were behind his back as if to hide something. The staff member made him show me what had been behind him. Heston slowly moved his hand to reveal an engineering masterpiece.

It was a small block of wood that had wires going everywhere. I couldn't tell what was coming and what was going. He had taken four plastic Pepsi caps and attached them to the side of the wood with nails. He had scavenged an old cell phone battery and had somehow rigged it to the plastic "tires."

I was very curious, so I asked him, "Does it work?" "Oh yes Papa, it works great." And with that, he bent down and attached a dangling wire to the battery. The wheels began to hiss and spin like crazy. He bent down and placed it on the floor, and it shot off like Grease Lightning. This thing was FAST! He ran after it and picked it up.

As he returned, the staff member asked me what I was going to do as punishment. I looked Heston in the eyes and said, "Congratulations! You are truly and engineer at heart. I will make sure you get everything necessary to help you. But what is that at the front of the car?"

"Oh, that is a nail so I can hurt my enemy if they come after me." I told him to remove the nail and only to use his engineering "powers" for good, not evil.

For Christmas, I bought him a science kit that allows you to do various experiments and projects.

His first project was creating an alarm system to detect if anyone enters his room. This kid is crazy. I honestly believe he will be one of India's brightest engineers one day.

South Africa Team

Last year we had the chance to host an impressive team from Global Challege Expeditions. This team of 11 had spent almost a full year travelling the globe. They had been to 10 countries before us, and we were the last stop.

The team was made up of 8 young college age adults and three young professionals. 10 of the members were from South Africa, and one member was from Brazil. They stayed with us for a week, and the boys had a blast. They shared diversity with us with African dances, and Lucas gave us a Brazilian influence and food.

The team fit in immediately, and it just felt natural having them wandering about. They taught the boys new songs and skits and even coordinated our annual Shiloh Olympics, or as we call it Shilohlympics.

I could tell that they were wary from the rest of their year-long journey. They had seen and experienced things most people will not in a lifetime. They were tired and just needed a place to rest and recoup along their journey.

During the day their time was filled with a myriad of activities with the boys. At night our staff got to spend more time with them, and I taught some how to play Yahtzee and introduced them to one of my favorite movies, Army of Darkness. One of my mottos is "Stay one night, and you are a guest. Stay two nights, and you become family."

This team also became part of our family because of their kindness and compassion. Their journey had been long, and many were your typical poor college student (haven't we all been there?). Their budget was bare bones. They had spent the year living and eating frugally, and their waistlines showed it. Many had lost an enormous amount of weight. But they weren't traveling the world in a culinary pursuit.

They were going to meet new people, and to fellowship, and to grow. So almost everywhere they went they ate incredibly simple. They told us their budget, and we took them to the market so they could buy their supplies for the week. The majority of it was rice. That first night I lay in bed and thought about their amazing journey. I thought about how awesome it would be if our boys were to take a similar journey. I thought of how grateful I would be if someone were to show extra love to my boys. I realized that we had that chance also.

This team consisted of other people's children. I thought I would pay it forward and treat these team members like I wanted someone to treat my boys. So we just fed the team as best as we could with no regard to their budget. We are also on extremely limited funds and couldn't go out and give them a five-course meal, but we fed them as best as we could. We did this because I know someone would do the same for our boys.

At the end of their time with us, we had one of the saddest farewells we have ever had.

We felt their absence for days. Their year-long journey ended soon after and they returned home. We are still in touch with the members through Facebook, and it is awesome to see how God is using their trip as a foundation for their life and future. We pray they will all come back and visit us sooner than later. And we praise God for the wonderful week we had with our South African team.

John Waller

In college one of my favorite groups was According to John. They got regular play on my Walkman. They had an excellent current sound similar to R.E.M. or The Waiting. I listened to them all the time. But that was the late 90's.

Fast forward to the beginning of 2016, and my friend Chris Leader told me that he was bringing a singer by our orphanage on the way to the Bible college graduation ceremony a few hours away. I was excited and asked them if I could catch a ride with them since I was required to be at the graduation anyway. They agreed that they could fit me in their van and we could all leave together.

In the weeks that led up to the trip, I still wasn't aware of who the singer was. Chris told me that it was John Waller. I admit that his name sounded familiar, but I wasn't exactly sure who it was at first. But I did what everyone would do in my situation; I cyber stalked him. Not really, but I did google him to figure out who he was. Since being in India for a few years, I have lost touch with the Christian music scene. I hadn't listened to much radio even before I left the US. But looking up some songs on Youtube, I realized that I knew ALOT of John's songs. Some of them were wildly popular on the radio and had been used in all of the Kendrick Brother's movies like Fireproof and War Room. I had heard his song about orphans and had loved it. I hadn't put a name with the songs, but I knew who John was. But it was only when Chris mentioned that John used to be the singer for According to John that I began to geek out. This was a band that I loved from my formative years, and I was excited to meet John face to face finally.

For someone who has tasted success, I preemptively assumed that he would be a little standoffish, a little conceited. I couldn't have been more wrong. John was approachable and humble. But most importantly, John was genuine. He wasn't fake, and he wasn't clamoring for the spotlight.

We didn't have a long time before we had to make the journey to the graduation, so I gave them a quick tour of the property. I had introduced the boys to all of John's albums and songs in the preceding weeks, so they were star struck. As we walked around, they would peek around the corner-wide-eyed at seeing someone who sang the songs that came out of my speakers.

After eating a small bite, John wanted to hang out with the boys. He asked if he could share a song and of course, I said yes. I was a little worried, though because in case you didn't know, we have a 30-foot dome in the middle of our building. The building was not built with sound quality in mind and when too many people sing at one time it reverberates into ear popping chaos. I was so worried that it would be too loud for John and his guitar. We were soon to find out. I made the boys sit down in front of our stage. We have a 3-foot platform where John would have been able to give a mini concert. But he didn't want to give a show. He wanted to have some quality time with my boys, so he got down on their level. That is right. John Waller got down on his hands and knees and then sat at eye level with the boys. It was funny when the first thing he told them was. "So this is what it means to sit "Indian style."" John sang the most beautiful version of his song "Our God Reigns Here." The sound was angelic. It was as if our building was built solely for that moment. From the guitar to the vocals, to the piercing words, it felt as if the song and the building were built for that very moment. It was a moment I will never forget. But it didn't end there. John continued playing for a solid 45 minutes; long past when we needed to leave. It was a time that I still cherish as do the boys. We will forever be grateful to John Waller, and the time he took with us that day. He made us feel special.

We left soon after, and John and I talked almost the whole 7 hours to our destination. I learned of his adoption stories. I learned of his nine children. I learned of his coffee company that was birthed out of his quiet times with Jesus. But most importantly I learned that John is a man of integrity who has a passion for not only orphans but people in general.

Recently John and his wife Josee have started Crazy Faith Coffee Company, http://www.crazyfaithcoffee.com/

It is phenomenal coffee that includes stories of crazy faith and also support orphans. I highly recommend checking them out.

From that fateful day last March until today, John and I have continued to grow our friendship. He even gave an endorsement to my book last year where he wrote,"I thought I knew what crazy faith was until I met Andy Lepper and his wife, Susan. What they are doing in India for orphan boys blew my mind. It is a lifestyle of faith that few of us have experienced. This powerful book will so inspire you. I am fortunate I got to witness firsthand the life of the man who wrote it."

Such amazing words from my now great friend John Waller. John is the real deal, and I encourage you to check out all of his music. By the time you read this you will probably already be familiar with his new song "Awakening," or as it has become to be known, "The Coffee Song." I pray that John will continue to be blessed in his music ministry and his business. We were blessed by his visit and intimate concert. Our boys will not too soon forget their personal hero, John Waller.

Boys Encourage Prayer Journal

As you may know, the first room that we fixed up upon "inheriting" the building four years ago was the prayer room. Well, to be fair we didn't renovate it. We created it. Previously there were books and desks just piled in the middle of the floor, and it had become a catch-all for household junk. The room is about 12 feet by 12 feet. We painted the walls and scrubbed the marble floors. But most importantly we hung a dozen or so dry erase boards on the wall to act as our Prayer list. It is still like that to this day. The boys enjoyed praying aloud for every name on that list. There is a good chance that many of you reading this book have had their names on our prayer list before. We love you and pray for you.

But the boys wanted something a more; a little more personal. I remembered that back in college some 20 years ago, I had purchased a prayer journal at a music festival. It was bare bones and straight forward. It had 500 entries for you to jot down all of your prayer request as well as space to log in God's answers. I still have that journal. I decided that I would create my own template so I could print it out and pass along to the boys. I printed out 100 or so sheets at a time and would staple 3-4 sheets together and pass them out. The boys loved having a prayer journal of their own. That was my only intention at the time; to help my boys in their prayer life. But seeing them use it so quickly and frequently got me to thinking about whether other people may enjoy this or not. So I got to work with my good buddy, Scott Cuzzo, and we created our own prayer journal and sent it off to be printed. Sharing these prayer journals has become a blessing for me. I enjoy nothing more than giving and its an honor to share this with others. We have made the prayer journal available to churches, Sunday school classes,

Bible studies, and anyone who wants one. Please feel free to ask us and we will send you one also

The journal has also become a blessing to other people because they have made it a part of their Bible study and prayer time. It has helped to focus their prayers and to give God the glory once they recognize how He answered the prayer. Like many things in my life, I can give credit to my boys. Their desire to know God more and to pray more helped facilitate the printing of this journal that has now blessed many.

At The Playground

We have given toys and gifts to the boys through the years, but we don't have many common things they can play with. We do always have Cricket bats and balls as well as soccer balls to play with, and there is usually a daily game of some sort. But we have not had a playground of our own.

Our next best option was to visit our local park and use their playground. We have done this a few times, and we all have a lot of fun. We pack a lunch and load up on the bus. We buy snacks and sodas on the way. It usually takes us about 15-20 minutes to get there, but we sing the whole way. Once we arrive, we make a beeline for their playground equipment. They have monkey bars, swings, slides, and everything you imagine a good old fashioned playground having. We usually wear ourselves out before sitting on benches or under a nearby tree to eat lunch. After lunch, the younger boys will go right back to the playground while the older boys will play a pickup game of soccer. Our boys always play hard. So it is not as safe for the smaller boys. None of our boys are overly competitive; they never play to win at all costs. But they are intense and only know one speed- full blast. So the younger boys hang back and prefer the playground when we are at the park.

It got me to thinking. When we are home, and the younger boys hang back, what do they play? They are very inventive and come up with some crazy made up games, but I wanted something more

So, for Christmas last year our goal was to build a starter playground for the younger boys to enjoy. It would also have swings for all ages to enjoy. We were easily able to fundraise the money and sought to buy a playground and have it set by Christmas. But things are not so easy in India. It took us a long time even to find a reputable dealer. There are zero big box stores in India like so many people are accustomed to. I couldn't just go 10 minutes down the street and buy a kit. Things here have to be ordered from overseas. And then the waiting begins.

We contacted many companies but most

never showed up to access our needs or space. Finally, in early February, we found the right man. And he even had a catalog! We picked out exactly what we wanted and waited three weeks for it to be shipped to him. Once he got it, it rained for a solid week, and he was unable to install. Finally, once he did install it, we realized that it was missing quite a few parts. So he ordered them, and we again started the waiting period. This time it didn't take as long because we urged him to drive to Delhi and pick up the missing pieces.

On March 20, 2016, we finally had the ribbon cutting ceremony for our new playground. The boys were excited. They each had a part in cutting the ribbon to inaugurate it. Our playground consists of 3 regular swings, one baby/toddler swing, two slides, a platform, a seesaw, merry go round, and spring horse. Our goal is to add to it every year.

Not a day goes by where almost all of our boys do not use some part of the playground. It has become the gathering place for free time, and now the boys even use it as a place to play cricket and soccer. Thank you so much to the ones who helped to fund our playground. It has been significantly used, and we truly appreciate your gift.

Goat Births

There is nothing quite like living on a farm. Some days it feels like the farm takes over and consumes us. But other days I truly forget that there is a farm outside my office door. With staff members helping to care for the daily needs of all of our animals, I can get lost in computer work and forget where I am. It doesn't take long for that fog to lift once I walk outside and get a whiff of our farm air. Take a few steps, and you are bound to step in various animal manure. But we all love this life and want nothing different.

For a long time, we didn't have our own male goat that was old enough to reproduce. We have had baby males born, but we usually would either sell them after a few months or grow them big enough to eat. (Goat curry is my favorite.) So how would we breed our females with no males that were old enough?

We came up with a brilliant but simple solution. We also do this for our water buffalos. We find a farmer near to us that has males. We invite them to come and bring their male animals to breed with our female animals. We then either offer them a small payment, (usually $1-2,) or we let them graze their males on our grassy fields for the rest of the day. So far they have all preferred letting their males free range on grass. So we can stud our females for nothing.

What this does, though, is cause a small window of breeding. All of our female goats are bred the same day. Which following logic means that their kids will all be born pretty close together. Of course, pregnancies are never the same from person to person or even goat to goat. We thought that maybe our four female goats would all give birth within a week of each other.

After a few months of pregnancy, we knew that the time was near. We prepared ourselves for the four births, hopefully, spread out over a few days. One morning we awoke to one of our goats giving birth. This is something awesome to watch, by the way.

I broadcasted it live on Periscope. It was funny that one person told me that I wasn't close enough and that I should move in for a better shot. Another person responded that they thought I was TOO close and should move back some. So I just split the difference and stayed right where I was.

As this goat was giving birth, all the other goats came around to see what all the fuss was about. No sooner did they all gather, did the second goat began to give birth. As she started into labor, the THIRD goat's water broke! We were laughing hysterically as we tried to get enough towels to assist with these three births. As soon as the first goat had started to push, and the feet were halfway out, the fourth goat went into labor. We had four different goats in labor at the same time!!!!! And it was all on video.

We were simultaneously laughing our heads off and running around trying to get things done. It is something I will not soon forget. When all was said and done we had six babies total because 2 of the mama's had given us two babies each.

My friend Lisa told me that when she was growing up, she had a super nanny goat that would give birth to four babies at a time. That is crazy. But I would prefer that to having 4 mama goats in labor at the same time.

We love our goats. They give us milk and meat. And they are very affectionate. They run through our building multiple times a day, with their bleating reverberating off our 30 foot dome roof. They bring life to our farm and home. We are thankful for what they provide for us.

Kanti

In 2015 Kanti came to personally help Susan. Susan had a high-risk pregnancy, so she needed someone to be her hands and feet for times when she was fatigued. Kanti had been a helper with another family in South India, but they were aging fast and no longer needed her.

Kanti was raised at an orphanage, and she has no family. She has no education, no training, and had nowhere else to go. She would have been put on the street had we not taken her in. I do not tell you this to act like we are saints for giving her shelter. Kanti is just one of many girls who are in the same situation. They have no one to care for them, no education, and nowhere to run. They are forced to be helpers in people's homes because they have no alternative.

Even from the beginning, we did not treat Kanti like she was a helper. We treated her like family; like a long lost sister. She wasn't hidden from site, toiling away in our kitchen. She went where Susan went; as a friend and confidant. In fact, she got the best accommodations we had to offer, our best guest room. She had more space than any of our other staff along with AC (she didn't want it), a water heater, and the softest bed we could find.

Once again, I am not trying to pat ourselves on the back. Just wanted to illustrate that we treated her as a guest or an equal more than a servant. In fact, she was more of a VIP to me because of the care she was giving Susan during her pregnancy.

But she was super shy and standoffish. She had never had much interaction with anyone and was confused when I attempted to carry on conversations with her. I think that was just never done where she previously was. It was more than six months before she even made eye contact with me when we spoke. That was tough for me. I was taught to look someone straight in the eye as a sign of respect when you talked with someone. It meant that you were talking with them, not at them. But I understood that it was not intended as disrespect on her part. She was probably scared to death that I would even interact with her anyway.

Slowly she became part of our family. She began to kid around and joke. More so with Susan, but she was finally opening up. I believe she felt like she finally had a family. From the beginning, Susan wanted more for Kanti than just to be our helper. Even from the beginning, we paid her a salary equal to what we pay our other staff. She was not expecting that and refused to take it. So we kept it in an envelope for her.

Susan's intention from the beginning was to give a better life to Kanti. In case you didn't know, the vast majority of marriages are still arranged here in India. It is not a Hindu thing. It is an Indian thing. Even Christians here arrange marriages. Susan worked for months to find the perfect man for Kanti to marry. After a few months, she found the perfect one. The arrangement, engagement, and marriage is a book in and of itself, but in the fall of 2015, Kanti got married. Her husband is a hard working man and has been a great provider. We have known him his whole life, and they make the best couple. We gave Kanti's salary that we had been saving plus some extra money as a gift for their wedding and seed money to start their life. She truly was a blessing, and we were blessed to help her.

It didn't take long before Kanti called us to tell us that she was also pregnant. In India, it is not uncommon for husband and wife to be away from each other for extended periods of time. This was true for Susan and me as we were separated for the first year of our marriage. Kanti asked us if she could stay with us after she got pregnant. This was not because of her marriage, but more to do with her husband constantly traveling and working. She didn't want to be alone. She wanted to help us and be close to Susan so she could get advice on her pregnancy.

So she moved back in. Susan went with her for all her pregnancy doctor appointments.

Kanti was still in the first trimester, so she was still at full speed. She was a huge blessing daily as she became the nanny for Micah. She was around him every waking moment. Even when we encouraged her to rest, she still wanted to hold Micah. I would watch her as she would be looking in Micah's eyes. I couldn't help but think that she was thinking forward to the day where she would have her baby. Very soon she will have that beautiful baby in her arms.

Once her pregnancy moved along, she returned home to be with her husband and prepare for the baby. By the time you read this Kanti will have a baby. It is illegal to know the gender of the baby before birth, so we do not know what she will have. You will have to send me an email if you want to know the truth :)

We have been blessed by Kanti. It has nothing to do with us, and more to do with just simple opportunity. Kanti came to us fragile, broken, lonely, and destitute. But today she is a strong wife, mother, friend. She is a strong woman. Not because of anything we have done. But because of the great God that allowed us to be there for her and provide her with a small opportunity. Kanti did the rest.

Jesus Toffee

Right before Christmas, we had a couple of friends stop in to see us. They were touring around and wanted to stop in. So we had a good visit from Larry and Mark. They had been here a couple of years before, but they were unsure if any of the boys remembered them or not.

I asked one of the boys if they recognized our friends and he said, "Oh yes Papa, these are the Jesus Toffee men." Jesus Toffee men? What in the world are you talking about??

We had no clue what they meant. When I asked him, he said that when the men were here before they handed out Jesus toffee. We still had no clue what he was talking about. He said that the men had given out candy hooks that were red and white. The red was for the blood of Jesus, and the white was for forgiveness.

At that moment we all belly laughed when we realized the boy was talking about candy canes. The trip before Larry and Mark had given the boys candy canes and told them what the stripes meant. Our boys never forget anything. Years later they still remember the story that Mark and Larry had told them, and they had not forgotten about the Jesus Toffee!

Badminton

Surprisingly, one of the favorite sports in India is Badminton. In fact, India has a professional league called the Premiere Badminton League. It is legit.

We have also loved playing badminton through the years. Our sports court outside has two huge poles where we used to run a volleyball net.

Until the fateful day that it broke. We haven't replaced it.

But usually every few months or so I will bring back a few rackets and shuttlecocks from the market. I did this a few months ago so the boys could just hit the birdie around a bit to let off some steam.

It was summertime and Susan was away visiting family. As I was sitting and watching the boys hit it here and there, I started taking a deeper look at the stripes that we had running through our marble floor. The markings divided our great room into 4 square sections. 2 of the sections also had the stage on it, so I knew they wouldn't work.

But I quickly googled the dimensions of a badminton court and was pleased to discover that half of our great room was almost perfectly marked off to be a badminton court. It was just a few inches short on the back lines.

A thought occurred to me, "When mommy is away, the boys will play." I knew that Susan would not let us have this much fun, so our time was limited.

I sent two boys off to bring back two metal beds. We stood each one, end on end, adjacent from each other to create the center line. We then strung a sturdy rope at the exact regulation height between the two beds. We made sure it was super taut because we then draped thin bed sheets over it to act as our net.

Once this was done, we had a jury-rigged regulation badminton court. We had the most fun over the next few days playing tournaments and kept meticulous scores. We would play matches where the winner stays, and the loser would rotate out. It was so much fun.

I had been running about 10 miles per day the weeks leading up to this and was keeping a good record of my steps on my fitness tracker. But I was so sore and tired from playing badminton all day that I didn't run for a few days. But the funny thing is, I got more steps playing than I did running.

We all had our duties to perform, but we would cover for each other and just play as much as we could. It was an excellent time.

Susan soon returned and quickly put an end to our badminton playing. She didn't think it was very dignified to be playing indoors during the daytime when we had so much other stuff to do. She was probably right. But it was so much fun.

We still joke about how much fun it was, and maybe one day soon Susan can take a trip so we can rekindle some old badminton rivalries.

Baptismal Pool

Sometimes you cannot give your kids what they want. Sometimes you struggle just to give them what they need. It's a hard road and tough balance providing either sometimes.

Last summer was a pure struggle for us. There were so many trials that we were going through that we were just being crushed from all sides. It was one of those feelings where it had gotten so bad that you thought that nothing could make it worse.

I cannot go into the details, but we were burdened. The boys felt it too. Kids can sense the stress their parents are under. No matter how hard we try, they still can sense it. Sometimes they feel like the cause. Sometimes they go out of their way to help relieve some of that stress from us. For me, knowing that my boys can sense my stress is almost more stressful than the actual stress. Does that make sense?

We had some tough times, but I just wanted the boys to feel like kids. They are not the source of our problems and the more that we can help them realize that, the better. We try to do little outings here and there like going to the park or eating out occasionally. But in the heat of the summer, it's hard to muster enough energy to do that.

When our building was constructed, they built a baptismal pool right outside our kitchen door. To the untrained eye, it just looks like an open concrete water storage tank. No water pipe is leading into it, and there is no drainage. But steps are leading down. Clearly, it's meant for baptisms. (Also it says so on the blueprints.)

When we reach our boiling point, I commission the boys to scrub the inside of the tank. They use bleach and brushes and make it sparkle. They know that this is just the precursor to a day in the water. After they have done their cleaning, we fill the water to about 3 feet. The tank itself is roughly 6 feet wide, 12 feet long and 6 feet deep. So adding 3 feet of water is a lot of water. But it is enough once you add 40 kids in there. The water level rises to what we like.

There are no pumps. There are no filters. There are only 40 sweaty, stinky boys jumping, dunking, and having the time of their life. I always get in with them, but only at the beginning. It doesn't take long before the pristine water becomes a murky brown. But the boys don't care. They play and play and play until their heart is content. And that is our intention.

I have approached a few places in our city that have swimming pools. I wanted to rent it out for any price so our boys could swim. I asked people to name their price. But no one will even give the boys a chance. Once they here that the boys are from our children's home the pool owners refuse. They do not want any dirty children in their pool. This is sad but true.

It truly doesn't matter. Our boys are valued and have value. Who needs to go to a hotel pool? We have our very own. We have a pool where we don't have to worry about etiquette. We don't have to worry about how clean or dirty we are. We just have to worry about having the most fun possible. And THAT is what it is all about!

Boundary Wall

Funding a children's home is tough enough. We are in a constant battle with having enough. There are a few things that we encounter that I just wish we didn't have to deal with. On the top of that list is our boundary wall.

We have roughly 3700 linear feet of boundary wall. It ranges from 6-7 feet tall across that 3700 feet. Suffice it to say we have ALOT of wall. This isn't fence that we can mend when it breaks. This is concrete plaster and huge stones. The stones were mined out of the nearby Aravalli Mountains. They were chipped down to manageable sizes.

The rocks are not our problem, though. It has become apparent to us that the concrete mortar mix holding the rocks in place is substandard. The wall is less than 15 years old but is in a constant state of disrepair. It is crumbling everywhere. The mortar has become like chalk and can be easily broken off and chipped away.

This may not be the biggest issue except for the fact that people have started using our wall as a walkway to get from one side to the other. This is not our boys. Our boys know well enough never to climb or walk on our wall. They do occasionally have to climb over it when one of our soccer or cricket balls goes over, but they know that other than that, the wall is off limits. But the people on the other side could care less about us or the wall.

The wall is on our side and was built during the time of our building. It is our wall. It separates us from open fields and abandoned plots. We have no issues with our neighbors. The children that play games on the other side of the wall are more to blame. They get curious and want to perch atop our wall looking down at us. The wall is not stable and is a hazard.

The last thing we want or need is to have a village kid fall off of our crumbling wall and get seriously injured. But they do not listen. They run along the top and have forced the top layer to come crumbling down. This has caused cracks all the way down, and we have had whole sections of the wall collapse. It is tough to keep a water buffalo inside our wall when she sees a field of fresh wheat through the open whole.

Along with the weak plaster, the foundation wasn't built very well. When the monsoons come yearly, we know that without a doubt, there will be sections of our wall that falls down somewhere. Last year we had to replace over 300 linear feet of our wall totally. The new wall was built better and hopefully will last a lot longer.

It will not be too long before all of the remaining wall will crumble and need to be replaced. Hopefully, it will fall in even increments so we can absorb the cost better. This is not something we plan for. We do not have a "boundary wall" budget. Maybe we should. So far we have been able to adjust the cost and make due. For that we are grateful.

It is also a testament to God. When things seem like more than we can handle, we turn to Him, and He comforts and provides.

One good thing that has come about in recent months. In the past, I encouraged our staff and boys to assist the mason when he would come to repair the walls. They were his labor, mixing the concrete as he guided, picking out the best stones, and just overall doing everything that he asked. They started observing to the point that a few months ago they told me that they were confident that they could do it as well as the mason we pay. So I gave them a shot.

And what do you know? They did a bang up job. It looks great. I fully believe that they did a better job than the expensive mason we were paying. It looks better, and I think it is more sturdy. We have positioned ourselves to be more sustainable by being able to fix our problems once they arise. Our boundary walls may always be a problem for us, but at least now we have a better solution to fix it.

We Grow Grass

This is not meant to be a double entendre even though I know it looks that way. I am not trying to be tongue in cheek when I say we grow grass.

Farmers for centuries have let their animals graze on grassy fields. We do not have that option to devote to letting the animals graze like that. Partly because we don't have a huge farm and we need to use our fields for staple crops like wheat, corn, and mustard. Another reason we don't let the animals free range all day is that our field is very deep.

If we were to let our goats graze on the back section, we believe there would be people who would jump the wall and steal our animals before we had the chance to run back there. Of course, no one is stealing a 1200 pound water buffalo, but there are still people in our area that would rather us close up shop than care for the boys. I wouldn't want someone to kill or poison our water buffalos when they were out of our site. It has happened before.

We just try to stay on top of things. Our original two water buffalos were stolen during our prayer time one night. Things happen.

So all that to say that we grow our grass for the animals. We give the animals a lot more than grass, so a whole field dedicated to grass is not needed. For our buffaloes, we feed them bajra stalks that have been ground up into chaff. We mix that with the grass we grow and add mustard oil. We cook it over an open fire outside to make it a bit more digestible. It may seem strange, but it's what they do around here.

The grass is an important part of their diet. The chaff is the equivalent to dried hay. The mustard oil helps the mothers produce more milk.

But we don't just give grass to the water buffaloes. Our goats and chickens also love it. Goats will eat anything but the love the grass. And there is nothing better this side of heaven than an egg that has been laid by grass fed chicken. It is out of this world.

The grass we grow is known locally as berseem. In English, it is known as Egyptian clover. Our animals love it. We plant an acre of it at a time. We have the field split into seven sections. Each day we go to a section of that field and hand cut all the beseem down to about 2 inches above the ground. This provides plenty of grass for all the animals. The next day we move to the next section of the field and repeat. By the time we get back to the first section the following week, it has grown back enough to give the same amount of grass. And the cycle continues.

When we first decided to grow our farm and get animals, I knew that I didn't want to bring in a lot of animal feed from the outside. I wanted to grow as much feed as possible. It not only saves money but it makes money. Take for instance the wheat we grow. If I fed 1 kilogram of wheat to our animals, I would get eight times more money from the sale of milk and eggs than I would if I just simply sold that 1 kg of wheat in the market. This is our strategy.

We grow our grass for multiple reasons- it is more healthy, more cost-effective, and actually, generates more money for us. We love our animals, and we love that we can grow our own grass to feed them.

S'mores

S'mores- that childhood classic. No childhood would be complete without the experience of sitting around a campfire roasting marshmallows. Couple that with being able to put the gooey marshmallows between two graham crackers and top it off with chocolate and you have, a little bit of heaven.

The last time I was in the US, a friend asked me if there were something that she could do that would be a little different than they are used to. Off the top of my head, I quickly said S'mores. Not sure where the thought came from, but that was what I was thinking. So my high school friend, Melinda Stembridge sent me a huge Amazon box filled with marshmallows, Hershey's candy bars, and graham crackers. This box was huge. It was so big I had to do some calculations and leave part of it behind. I just didn't have enough room in my suitcase for 30 pounds of S'mores. But I was able to pack enough for everyone in our home to eat at least two.

When I arrived home, I told the boys that I had a surprise for them. I told them that we would build a massive fire and make cookies. I waited for them to finish their evening prayer. As I waited, I stacked enough wood to build a small house in the form of a bonfire. This was going to be an epic fire. I put cardboard boxes at the very bottom, and I was ready to go.

I assembled the boys outside once the sun went down and we started the fire. It was blazing hot in no time. I instructed the boys to get thin, sturdy sticks that we could use in the fire. We don't have enough coat hangers to use for the marshmallow rods, unfortunately. The boys totally misunderstood me because they all ran back with their perfect sticks and began throwing them in the fire. I told them not to throw them in the fire again, and they ran off to get new sticks.

Once they had the sticks, I gave them each a piece of marshmallow and told them to roast it. I ran inside and began to assemble the chocolate atop the graham crackers. They were soon done with their marshmallows and began to line up in the doorway as we helped them assemble their S'mores. They had a blast. Everyone loved it. We had enough ingredients, so we went ahead with round two.

All of the boys had a blast, and even the staff had an excellent time. The best line of the night was when one of the smallest boys approached me. I asked him if he had a good time. His response was, "Oh yes Papa. I am excited for the next time that we can make "Fire Biscuits."" And there you have it. The next time you enjoy a S'more, think about our boys, and all the fun they had eating fire biscuits.

Rainstorm

Late one typical afternoon, with no warning, the skies turned gray and ominous. The wind began to pick up, and we knew we had a fast thunderstorm blowing through. Our city of a million doesn't have a bunch of news channels. In fact, there is no TV station at all. So we never get our local weather. Occasionally I may look on the internet, but the forecast rarely changes. It usually just says HOT.

The random intense thunderstorm is rare. But with the conditions changing as fast as they were, we knew we were in for a rough go of it very soon. Within moments the winds became fierce, and rain started falling. Rain started coming at us sideways and right behind that was the hail.

It was raining golf ball size hail in the midst of the torrential downpour. Our ground was so parched and dry that it didn't take the water very well. It just created a flash flood right outside of our back door. We could see a few trees falling over in the wind. The tin roof of our chicken coop blew away. Our neighbor's satellite dish ended up on our side of the wall. Our whole water buffalo house toppled over.

It was a fierce storm for about 30 minutes. And just as fiercely as it came, it was gone. It went from being almost pitch black to the sunshine in full effect within a matter of 4 minutes. It was crazy to watch and experience.

It happened so fast that we were not prepared. We had our buffalos tied to a tree, and our first thought was to rush out and move them back to their hut. But we were afraid that because of the flooding that someone would get injured. Thankfully we did not move them because shortly after their hut collapsed and they would have been injured or in the very least freaked out.

Once things blew over it was time to access the damages. Other than heavy flooding and stuff that could be somewhat easily repaired, we didn't sustain much damage. But unfortunately, with the flash flooding, we had a chicken that drowned.

When we found him, the rooster was barely

holding on to life. His eyes were already glazed, and he was barely breathing. I thought the best thing we could do was put the poor chicken out of his misery.

But nothing goes to waste here, so I got the boys to butcher him. Since it was only one chicken, he didn't provide enough meat for everyone, but we were able to make an amazing chicken soup full of carrots, potatoes and all the other fresh veggies we had on hand.

This became a life lesson for a few of the boys. We had not intended to eat chicken that night. But we had to make a tough choice in culling the chicken. Instead of just throwing it away, we were able to eat it. When life gives you lemons, you make lemonade. Or in our case; if life kills your chicken, you make chicken soup.

Hot Night in the City

This is not a story with a happy ending. But it is not a story where we are sharing because we want sympathy either. Our conditions are rough here. The power goes out about 20 times a day. That makes it impossible to heat during the winter and cool in the summer.

Truly, all we desire in the summer is just moving air. We just want to be able to have electricity to power our ceiling fans, so the air doesn't stagnate. In the dead of summer, our temperatures can reach 120 degrees or higher. It is a dry heat, but that is little consolation.

We just want to feel some wind, even if it is hot. Many times I have had to drive my motorcycle during the hottest part of the day. This is not by choice, but sometimes duty calls. It the heat is almost unbearable. No matter how fast I drive, it doesn't improve. It feels like someone is standing in front of me pointing an industrial hair dryer at me with the setting on super high heat. It will take your breath away. But crazy as it is, I would prefer hot air to no air during the middle of the day.

On to the point of the story. One fateful night our power went out shortly before 7 pm. It was a viciously hot night. It was sticky and humid. There was nowhere anyone could go to escape. Not even the chickens.

We try to do the best we can with our chickens. Their chicken coop is made of brick and plaster. It has a nice aluminum roof. We have even hooked up lights and electricity. We have big windows for the air to flow, and two fans on adjacent sides to help circulate the air.

But on that night, with the power out, the fans didn't work. The chickens had no air to circulate. Even the curtains over their windows had fallen, trapping the hot air inside. It must have been miserable.

The power did not come on until late the next day. But in the morning before it came on, we went to let the chickens out like we do every morning. They must have had a miserable night because we found a total of 18 chickens had died during the night. 18! At the time we only had about 100. And all 18 were hens who laid eggs regularly. So not only did they die, but we lost our main egg producers.

It was pretty sad. We had no way of knowing when during the night they had died, and their bodies didn't look too good when we found them, so we were forced not to eat them. And we couldn't just throw them away because we may attract unwanted dogs.

We had no choice but to burn them. It was a sad day. There have been many days we count as victories with our chickens- the days they feed us with meat and eggs. But this day was a major loss. We pray we will never experience a day where we lose that many animals at once.

Weather

Many people might classify me as crazy, but I am a lover of all weather. I enjoy every season. I love the cold temperatures of winter and the hot hot heat of summer. I grew up in Beaufort South Carolina, right on the Atlantic Ocean, and summer to me means the beach. But I have had my fair share of cold cities too. But I have never encountered weather like we have in India.

Weather is always a factor to us. If you look up our city on Wikipedia, it will tell you that for 60 years our city had the highest recorded temperature in India. It was only broken last summer.

It is hot here. We are at the base of the Aravalli mountains, the oldest mountains in India.

We are also at the beginning of the Rajasthan Desert. Our climate is closely related to Arizona or Egypt. People assume that we only have a hot, dry climate. While we do have that, it is not the full truth.

As I write this in early March, the temperatures have already reached 100 degrees. May is our hottest month, and we will see temperatures close to 120 degrees this summer with very little humidity. But that is not the extent of our weather.

Once the monsoon comes in late July, we will go from the desert like conditions to more tropical conditions. I have been to Haiti over a dozen times, and that is what it feels like during our monsoon. In late July/early August, the temperatures will "dip" to 100 degrees, but it will bring 100% humidity. We go from one summer extreme to the other.

But the good news is that it is not always hot. We actually have mild and cool winters. It is mostly dry, but December and January we have temperatures that can dip in the high 30s. On occasion, we even drop below freezing. But there is never really a chance of precipitation.

The best time to visit us is either February and March or October and November. These are the mildest times. So, from freezing to temperatures that are the hottest ever recorded in India, our weather patterns fluctuate. You never know what may happen. But plan on the heat.

Blessing and a Curse

Our building is both a blessing and a curse. What a blessing it is to have such a large shelter. We can do so much with the building we have. There are about 30 rooms not counting small rooms. We have ten bathrooms. We have beautiful marble everywhere. We have a 30-foot cathedral ceiling. Our land is spacious. One of the rooms of the building was built to be a medical clinic. How cool is that? We have a huge marble stage as you walk in the from the front door. We have four private guest rooms. It doesn't take much to see how much of a blessing this building is. It is a gift.

But it is also a curse to us daily. This diatribe is not meant to show ungratefulness. We are blessed to have a roof, but I just wanted to share some of the idiosyncrasies of this building. It is both beautiful and weird.

When the building was first constructed the foundation was not as thick as it should have been. It should have been at least 1 foot thicker. Because of this, our marble floor flows seamlessly into our driveway. There are no steps or even a gentle decline. Its just straight into our dusty front driveway. We live in a desert climate, and it is dry and dusty most of the year. Because our foundation is so thin and low, the dust just blows from the highway straight through our front door. Our boys and staff spend about five man hours a day sweeping up this dust that flows inside. Five hours. Every single day. Because the foundation was not done properly, the building occasionally shifts, creating super long cracks in our walls.

The electrical work was not installed properly, to begin with, and was subpar at that. When I first arrived there was only one light that worked and two ceiling fans out of the 30 rooms. One time I went into a room that had an uncorking ceiling fan. I thought that the motor might be frozen, so I got a stick and pushed on the blades. Sure enough, it did start moving. But it didn't stop. It just kept spinning faster and faster. As the speed increased, the shakiness and creaking sound also increased. It sounded like the Titanic when it was sinking. It was moaning and groaning. Before I could turn off the power switch, the fan exploded and sent one of the metal blades shooting through the glass window in the opposite direction. Thankful I was able to get out before a fan blade flew in my direction. Not long after that, we had the whole building rewired.

In our four years here we have spent well over $100,000 in upkeep and renovations. I feel like Tom Hanks in the movie Money Pit. We have had to totally re-concrete our whole roof because we had over 50 leaks at any given time. We have fixed guest rooms, and renovated the boy's rooms, and so much more. Every time we turn around it seems that something else has broken and needs replacing.

The dome on the top of our building is a local landmark. People know exactly where we are when we tell them to look for the dome. But unfortunately, it has an adverse effect on us. Because heat rises, heat accumulates in the top of the dome during the day. It will be 115 degrees outside, and that hot air rises to the top. Later at night it gets cooler outside and feels pretty good. It may be in the high 70's. But that is the time when the heat from the dome begins to ascend. So even though it is nice and cool outside, it is over 100 degrees inside our building at all times. We have never been able to come up with a proper solution effectively. Ceiling fans are inadequate as is everything else.

It is true that this building feels like a curse at times. But it is our building. It is our home. There is nowhere else that we can all be under one roof like this. We worship in this building. We eat, pray, study, work, sleep; you name it. It is our home, and no matter how bad it gets, it is still more of a blessing to us.

Dinosaur Uncle

We have a boy name Raj. This kid is amazing. He is ten years old. He reminds me a lot of myself when I was his age. He is compassionate but forgetful. Caring, but a little ADD.

One day he got on the bus and wasn't wearing his shoes. He honestly didn't even realize that he wasn't wearing shoes because he likes to go barefoot around the home anyway. He searched and searched and couldn't find his shoes. The bus left without him. After a few hours of looking, he finally found his shoes in his book bag. He looked at me, and you could tell that he just remembered putting them there. He told me that he didn't want to wear the shoes on the bus, so he put them in his book bag and was going to put them on once he got there. He is so forgetful.

But man is he a boy after my own heart. He is messy, forgetful, and mischievous. Recently the boys drain clogged in their bathroom. The pipes are huge, so we couldn't figure out what had caused it. A couple of boys came to Susan and confided in her that Raj had stuffed some toys down the drain. When asked why he did it, he replied that he just wanted to see if they could swim.

He is also tenderhearted. One day I was hanging out with some boys, and they were calling Raj by a nickname I hadn't heard him called before. They were calling him Dinosaur Uncle. I had no idea why. I asked the boys why they called him such a weird name. They told me it was because Raj liked to keep all kinds of animals under his bed. He was like the guy in the Jurassic Park movie, so they called him Dinosaur Uncle. I told them they must be mistaken because Raj would never keep anything under his bed. They told me to look with them. Raj had a big sheepish grin on his face when he realized that I was about to discover what he had been hiding.

Normally the boys keep all their belongings under their beds in their metal foot lockers. Some will keep soccer and cricket balls also. Raj didn't have any of that. He had put his foot locker with his clothes under someone else bed. Under his bed was a menagerie of animals. He had his own little zoo going on. He could have charged admission from the village kids.

Under his bed were clear bottles with an ant farm, other bugs, a baby bird in a shoe box, a lizard, a frog, and an injured ground squirrel. You read that right. He was keeping a full grown squirrel under his bed. You gotta love this kid. He was a little embarrassed and scared that I would remove it all. But I bent down and told him that it was ok as long as he helped the animals to live. Don't kill or harm any animal, I told him.

I also told him that I wouldn't tell Mommy if he didn't. So far, she hasn't found out about Raj's zoo. And she doesn't know the true meaning behind why we call him Dinosaur Uncle.

Extreme Flowers

We have very extreme weather here. We have the hottest summers and some pretty cold winters. But we also have resilient flowers. Near our front gate is a guard shack. It is a brick building barely 6 feet by 8 feet. Its primary function is to house our electrical grid that we run into our main building. There are no windows and no doors. It doesn't get your attention when you walk or drive by it.

But what does get your attention is the stone wall that is just behind it. Because on that wall there are the most beautiful flowers in the world. They are my favorite flowers in the world. They are my favorite for three reasons. And these three reasons are also life lessons for me

No one planted them. They just started growing one day. They didn't start life with fanfare. No one carefully tilled the ground and meticulously planted them. They just grew where they happened to be.

The conditions are harsh. How does such a beautiful plant grow in weather and soil so rough? Not a lot grows in our desert. But they are resilient and refuse to die. No one waters them. They have learned to adapt because they had to for survival. They know where to get the water to survive.

Looking at these flowers remind me to

1) grow where I am planted.

2) endure harsh conditions

3) learn to adapt.

When you look upon the beauty of a flower like we have, you see past the hardships it took to grow. You just marvel at their beauty. This is a lesson I try to pass on to my boys. One day I will not be there to "water" them. One day the situations they are in will be rough. And one day they will be in a place they didn't expect to be. How will they respond? Will they wilt and die? Or will they bloom despite the harshness around them?

The Weight of Struggles

It's not the weight of your struggle but how you carry it. This thought hit me like a brick in the face this past week. I was heading into town on my motorcycle as I do almost daily. It is a split road with a fence and some flowers in between the four lanes of traffic. My two lanes into town are not very wide. Passing isn't the easiest task on the best of days but it doable.

This was about 6 am. Otherwise, there would be bikes, cows, pigs, people, and cars galore.

And then I came upon him. An old man on a motorcycle that was visible struggling. He was carrying over 20 bamboo poles on his motorcycle each at least 10 feet long. First, let me say that I know how this guy may feel because I have firsthand knowledge. Not too long ago we were building a chicken coop, and I bought 30 of the same poles.

But the main difference is that I had someone to help me. Someone sat behind me on the bike and balanced the massive poles on their legs as the poles pointed straight ahead and behind. I cannot imagine ever thinking that balancing them on the back would be an option. At the very least you need someone there to help.

And that, my friends, is the point. If you are carrying the weight of your struggles by yourself, there is nothing bracing you from a fall. Someone to share the weight and balance it as you steer is crucial.

The Bible also speaks to this. 1 Peter 5:7 says, "Give all your worries and cares to God, for he cares about you." What is the point of doing it all yourself if you crash and burn? God cares for us. Give him what you are struggling to carry.

There is a saying that goes something like, "God will never give you more than you can handle." That sounds all warm and fuzzy, but I greatly disagree with it. If we could always handle what comes our way we would never have a need for a Saviour, and we would never glorify Him when He rescues us. The truth is all He does is give us more than we can handle This way when the weight of our struggles become too much we will turn to Him.

The truth is, if I can be transparent and vulnerable with you, I am feeling the weight of my struggles right now more than ever before. I am overwhelmed here at the children's home. We have very little support, and my family's personal savings are long gone. I am on fumes, and I cant do it anymore. I am ready to throw in the towel and admit defeat.

But this is precisely the time that I need to crawl into my savior's lap and trust that He has this. I am so exhausted I just need rest. In a sense, the poles on the back of my bike are about to make me crash and burn. But I know that I will never find rest in myself. No amount of free time or sleep can give me genuine rest. That only comes from God.

What struggles are you dealing with? How intense is that weight? The sooner you realize that God has your back and He is the only one who can handle this, the sooner you can give Him glory for who He is.

And finally, I waited around to see how the old man on the motorcycle would do. No more than 1 minute after the photo was snapped, 2 "random" men rode up in between us. Just from body language, I surmised that they probably didn't know each other. The guy on the back of the bike got onto the back of the old man's bike as the new driver placed the poles strategically on his leg. And then they were off. I am convinced that the old man didn't know the guy that helped him. But there he was sharing the burden and carrying the full brunt of the weight of the struggle.

God is approaching you and me also. Will we cast our burdens on Him and let Him carry our struggles? True rest is found there.

Arjun

I stood on the roof overlooking the boys at play. I noticed two boys in particular. Arjun, and Myron.

When Arjun arrived, he didn't know how to interact with other boys or anyone really for that matter. He came from a village in the jungle, and he just didn't have too many human interactions.

He came with Reeves. Reeves from day one assimilated well. In no time he was a part of everyone. But Arjun took so much longer. He has been here two years.

At first, when the boys had playtime or would watch a movie, Arjun was nowhere to be found. Many times I found him sitting on his bed or in an empty room just hanging out.

Don't get me wrong, though; he wasn't unhappy. He was perfectly happy being by himself. And that is ok. There is nothing wrong with that. But it's hard to live in community when everything is going on around you, and you just don't know how to adjust to that.

I remember countless times at the beginning where I would make it my objective to find Arjun during community free time and either hang out with him or invite him to play with the others. But I never pressured him. Most of the time he and I would sit on the low-lying concrete wall and watch the other boys play. And he was ok with that. He is not antisocial; he was just acclimating at his speed.

Before you know it Arjun would come and observe without my invitation. There was no longer a need for me to make him feel a part. He was joining in at his speed.

Fast forward to today, and as I walked toward the edge of the roof, I could hear the boys belly laughing. I wasn't close enough to see them yet because our roof is huge, but I could hear them, and one laugh in particular.

Let me be honest; I know most of the boy's laughs by heart. Our power goes out frequently, and I have grown to memorize their laughs, their voices, and even their shadows and silhouettes.

But I could recall this laugh. As I peered over the edge, I could see that it was Arjun that was leading the laughter. I had never heard him laugh like this. It was a deep belly laugh that you have with only those most intimate with you.

Arjun was standing in a crowd of 7 other boys, and they were playing a game. I am not sure what made them laugh, but they thought it was hoot-nanny hilarious. After a while the game restarted and they played and ran and dodged the tennis ball they were hurling at one another.

There stood Arjun shouting out orders to his team. He was dictating where they ran, when to jump, and when to chase. And then it hit me. Not only had Arjun overcome his shyness and become a part of his small group of friends; Arjun was their LEADER!

Just writing this gives me endless hope and joy. This is why we do what we do. We are in it for the long haul. NLO is not a fly by night organization that is in it for the glory or awards. We are in it to see these steady and slow tangible outcomes.

One day Arjun will be a husband and father and leader. He may not remember the time we invested patiently holding his hand and praying that he would come out of his shell. But we will remember, and memories and victories like this are why we do what we do. It gives us the fuel to keep pressing on.

What are the Small Things Worth

If you didn't know, we live in India. I see some pretty amazing things. Some of them make me laugh. Some of them make me cry. Some of them make me think.

Today was a thinking kind of day. I encountered a few things worth blogging about. I will share one experience with you.

I was having a nice leisurely ride on my motorcycle when I passed him. At first, I didn't catch what he was doing. He was slumped over on his hands and knees on the asphalt picking something up. That something was grains of wheat.

As I became aware of what he was doing, I got a wider look. And don't judge me, but I stopped to watch and didn't help. About 20 feet ahead was his motorcycle with a 50-pound bag of wheat on it. Something had happened, and the bag had lost some of its content. But not much. Maybe a pound of grain at the most.

But here was this man on his hands and knees picking up one pound of wheat that was spread out over 20 feet. I repeat, On his hands and knees. WHY? Was he so poor that this was his family's rations for the month and that 1 pound meant a day that his children wouldn't eat? Or was he afraid of what his boss might say if he found out a mistake had happened? I will never know the reason he was so intent on rescuing a pound of wheat, one grain at a time.

But what strikes me the most is his intensity for such a little thing. The truth is that he probably could ration the remainder and stretch it, so the loss is minimal to his family. The truth is also that his boss probably would never actually know that some were missing if that was the reason.

This smallest of things mattered to him. You could tell that each grain had value to him.

What are the small things worth to you? Even if no one would ever find out, do you value the insignificant things like this Indian man?

Luke 16:10 One who is faithful in a very little is also faithful in much, and one who is dishonest in a very little is also dishonest in much.

The Goat and the Rope

One day I was in my office, and I heard a struggle outside of my window. I honestly did not know what it was. There were no sounds of a struggle; it just didn't sound normal. I wasn't too rushed, but I got up to walk outside.

As I did, a staff member rushed passed me heading through our big kitchen. I quickly followed. We made our way out the door to find two goats in a deathly embrace.

All of our goats have posts that we tie them to for a couple of hours a day. This is in the shade, and they have access to grass and water. We do this so staff can eat or anything else they may need to do without worrying about a goat running away or getting stolen.

Our oldest male goat has huge horns. As we ran outside, we noticed that he had his horns stuck in the rope collar of a smaller male goat. The bigger, older goat wasn't hurt but more than anything he just wanted to get free. But the more he struggled to get free, the tighter the strain was on the smaller goats neck.

By the time we got to them, the smaller goat had gone limp. By now, every staff member had run out and crowded around. I shouted for someone to get me a knife. No one wanted to use a knife. Someone said that the rope was expensive and that we shouldn't cut it. I quickly ran inside and got the knife myself.

As I ran back, a couple of people were carefully trying to loosen the rope so it wouldn't fray. I was shocked. They were so nonchalant in trying to save the rope that they had forgotten the limp goat dangling from the horns of the big goat.

I pushed everyone aside and told them that I would buy a new rope but that we only had precious seconds to free this goat before he would die. I quickly cut the rope and threw it aside. The goat tumbled to the ground. I cannot say that I did

CPR on our goat, but I would have had I needed to. We poured cold water on his face and down his throat. I did chest compressions.

After a few moments, he came to. He was weary at first but was fine after a few hours. There have been no lasting effects.

The point of this story to me is not to make the staff look bad, or to make myself look better. The point of the story to me is that sometimes we put all of our focus on things that don't matter. We end up killing our goats because we want to save our ropes. It doesn't matter how nice the rope is. The rope only cost $2 here. The goat was $100. Focus on the goat. As simple as a lesson can be- let us focus on what matters, not on things that don't. We will have the life choked out of us if we concentrate on the insignificant things.

Filling up the Love Tank

There are a few vlogs I watch on Youtube. If you are not familiar with what a vlog is, it is a daily video diary, a video log, hence the name VLOG. My all time favorite vlog is pjoshyb & familee. They release daily vlogs about their life and ministry to students in the Northwest United States. Their videos are usually around 10 minutes, and I haven't missed a day since I discovered their Youtube channel. It is lighthearted fun and riveting. Watching their daily videos blesses me, and it is a part of my everyday routine. They have become friends in real life, and I encourage you to check them out.

I have another friend that also has an amazing Youtube channel. His name is Justin Rhodes, and he also vlogs daily. He is known as the Permaculture Chicken Ninja. His channel has centered around using permaculture to help you be as sustainable as possible. He and his family live on a small homestead in the mountains of North Carolina. As of the writing of this book, Justin and his family have renovated a school bus into a home on wheels, and are spending the year traveling around visiting like-minded homesteaders. He calls it the Great American Farm Tour. Their family is amazing, and I would also recommend checking them out.

Justin has a young son around two years old that they have affectionately nicknamed "Mr. Brown." He was nicknamed after the Dr. Seuss book. In one of Justin's daily vlogs, Mr. Brown was running around Justin's feet as he was outside doing farm chores. Mr. Brown would run a few feet away but then run back and grab Justin's leg. Justin made a simple comment, but it has stuck with me. He said, "Mr. Brown, do you need me to fill up your love tank?" And with that simple question, Mr. Brown was off again. Mr. Brown needed a moment of encouragement; a moment of love; a moment to know that Justin was still there. This, of course, is not a new concept. But it hit me in the right spot at the right time.

Being a new father of a child who is getting close to Mr. Brown's age, I can see the importance of this love tank. I "get it." As parents, we are called to give our children both roots and wings. As my son Micah begins to explore, he knows that I have given him the wings he needs to explore this newfound freedom. But he also knows that his roots have been laid down. He knows were to return to get his love tank refilled.

Like a typical toddler, one moment Micah is pushing me away. And the next moment he wants to be held. The fun comes when he simultaneously wants to be held and pushes me away. Isn't this a metaphor for life? Do we not run ourselves ragged and neglect our own Father because we are exercising our "freedom?" But because of the roots of our faith, we know where to go to fill our love tank. I will not try to overcomplicate something so very simple.

As with almost every life lesson God gives me, He points it to my 30 boys also. This is where my real learning comes. As I try to lay foundation and roots for Micah, I cannot help but think about all my other boys. What roots were they given before they came? What foundation? The fact that they are even here in the first place points to a broken foundation; of shallow roots. It doesn't matter the intentions or lack thereof; they come from broken homes.

So where do they go when they want to have their love tank filled? Are they aware that they can even do this? Or have they been taught that love is a weakness, something we must never mention? I have found that most of my boys do not know how to receive love or affection when they first come to us. They want it and need it; They just do not know how to receive it.

But we know that love is patient, and love is kind. Love is not forceful. You cannot force roots on someone. You cannot quickly build a strong foundation. It takes time. Our boys are beginning to understand the foundation we are trying to provide, the roots we offer. And they know the love that is theirs for the taking. And when they are ready to fill up their love tanks, we will be here.

Welcome to the Family

One day I found Joshua staring at a body building magazine. He read all the articles and saw all the pictures and was convinced that he wanted to have the body of a body builder. He begged me to buy him protein powder and supplements like the ones that filled the pages of the magazine. I had some protein powder already, so I just gave it to him and taught him how to make a shake.

A couple of weeks passed, and I asked him about his goal. He told me that he was taking the shakes daily, but he didn't see a change. I asked him how much time he was spending lifting weights and he told me he wasn't lifting at all. I looked at him like he was crazy. If you want to be a bodybuilder, you must BUILD your BODY. He told me that he wanted to be the first bodybuilder ever that didn't lift weights or exercise. I said that that is not how life works. You will not see results if you do not put in the work.

Caring ABOUT something is not the same as caring FOR something. If you care for something, then your life reflects it. Almost everyone cares ABOUT orphans. I mean, you would have to be a pretty miserable person not to care about the needy or at risk. But do we care FOR orphans? God did not call us only to have a regard for orphans; not just positive thoughts or good vibes. He never asked us to only pray for the orphan. He called us to go; to visit, to love, to weep, to supply. Of course we are not the only organization that cares for these children. So we are not assuming that people do nothing if they are not supporting us. We are not in competition with other organizations. We are all in this together and we pray for all of our fellow workers. We just want to encourage everyone to help meet the needs of the needy if they are not already doing that.

We have an amazing group of friends who do just that. They not only devote their prayers for our boys, but they stand in the gap financially. They send letters and videos to the boys. They provide Christmas, birthdays, and daily survival.

We are not obsessed with money. Never have been, never will be. We do not long for great and outrageous donations. We yearn for relationships to grow with our partners. We have settled on a motto for No Longer Orphans–WELCOME TO THE FAMILY. This is important to us for two reasons. First, when we take these boys in, we welcome them to our family. For some boys, we are the only family they have. Second, when we partner with you, we also welcome you into our family.

I am sincere when I say that we don't want people just financially to support us. We want you to become our family. Donors will come and go. But we need family to lean on during good times and bad.

We have people visit us all the time. Maybe you the reader can visit us soon. I always tell everyone when they come that the first night they are our guests. But the second night you are family. And because you are our family, money is never a factor. We have never charged anyone to stay with us. We will give you the best we have to offer. A warm bed, a hot shower, a good meal, and fellowship. If we visited you, would you not do the same? If I stayed with you how much would you charge me for the night? Of course, nothing because you invited me into your home.

Likewise, when people visit us, they are not visiting an orphanage. And honestly, it is not even a children's home. More than anything this is a family home. We have love enough to welcome as many people as possible into our family. The boys need the love, and we need the shoulder's to lean on. So I encourage you to not just care about our boys. Join us and care for our boys. We need you. Welcome to the Family!

What Does 2017 Hold?

Truthfully, we are not sure what the immediate future holds. Recently a prominent Christian organization was asked to leave India. We will keep doing what we are doing, for as long as we can. We have hopes and dreams for this year, and we are praying that God will make way for us.

One of the biggest things we would like to accomplish this year is to break ground for the building of our school by the end of the year. We are nowhere close to having full funding to build the school, but we are blessed with more than enough land for the school. Our desire is to give our boys the best education we possibly can. To do this, we need to have more control on how they are taught, so they are our first priority. But a huge factor to also consider is that we want to open up the school for nearby children. There is not an affordable school anywhere close to us, and there are no public schools for many miles. We want to be able to offer the best education for the least cost. One of the great benefits of building a school is that it will help make us almost 100% locally sustainable. Building a private school and opening it to outside students will give us the funds to be able to grow our children's home and to take in more kids.

One of our other major goals this year is to be able to branch out and to help other struggling children's homes. We want to be able to come alongside the home's directors and help them in their greatest time of need. But it will not merely be a hand out. We want to be able to pass along the knowledge we have of farming and animals and handicrafts so they can be as sustainable as possible themselves. Our goal is to join with them for a short time and equip them to care for themselves. Our goal is to phase out their need for support within 3 years. We have a list of homes that we want to help. We just need to see how God provides

Our last big initiative for 2017 is to be able to use our farm as a incubator for others to raise animals. We want to be able to gift our animals to others who are in need so they can provide for themselves like we do. We will start a breeding program on our land to be able to gift these animals to people in need.

Extra Acknowledgments

We want to thank the following donors and friends by name. These friends have financially supported us this last year and we are grateful. We have done the best we could with the list. There may be a few names missing but it is unintentional. We are truly sorry if we forgot to include your name.

Thomas Cecil, New Horizons Church, Julie Joseph, Schlieman Family, Brian Nunes, White Creek Baptist Church, HeidiJohnson, The Baptist Church of Beaufort, Wesley Denton, Mark Schama, Michael Pope, Cindy Stoutenborough, First Baptist of Lake Park, Caro Robertson, Doug Jinks, Loretta Dukes, Jason Johnson, Erin Sprinkle, Buckhorn United Methodist Church, Kelly Buchanan, First Presbyterian Church of Greenfield TN, Caroline Critzer, Jaime Adams Kiokee Baptist Church, Luke Mechtly, Marcia Horner, Marilyn Morrow, Olivia Brown, Scott Koenig, Gates Of Praise Church of God, Mona Loyd, Jeff Dolan, Carrie Whaley, Jay Hutchens, Stephen & Tracie Renfrow, Guy Spillers, John Bengier, Rick Tomlinson, Charlene Jenkins, David Bissette, Michelle Parnell, Amanda Raines, Amy McCarley, Kiokee WMU, Janet Johnson
Kevin Kyne, Mike Connelly, Brian Holbrook, Kelly Bryant, Al Lepper, Chris Leader, Eleanor Cox, Mary

Gowland, Katherine Pritchett, CornerstoneTile and Granite, LLC, Carole Courcoux-Allyn, Heather Hahn, Chris Fischer, Andrew Discavage, Danielle Kinney, Jena Holbrook, Sidney Bass, Jennifer Allison, Kaleena Burnett, Christie Ketterman, Josep Cato, Melissa Thomas, Claudia Glasco, Ross Jenkins, Patrick Henry, Crystal Saunders, Gresham Brown, Rebekah Wilson, Mary Holmes, Laura E Swango, Leslie Bumgarner, Jolly John, Dana Kendrick, Charla Archie, James Cooper, Kyle Brown, Brandy Spillers, Marcia King, Marcia Bailey, Elizabeth Williams, Laurel Lindner, S Green, Wendy Peterson, Anita Curtis, Casey Engen, Heather Peacock, Shelley Cooper, Darcy Boerio, Eric Goodwin, Sue Tuten, Dellila Hodgson, Amanda Lepper, Bethany Singleton, Jacob Kallara, Susan Dickey, Timothy Metz, Sue Fallin SharonLoar,CharlesCoble,Nancy Biggs,Barbara Akin, David Dent, Sparked Living: Life Purpose Coaching, LLC, Melinda Gordon, Keila Monroe, Cara Henry, Christine Hawkins, Marisa Hornish, Gerri Hanson, Ruth Campbell, Melvin Brown, Danae Medlin, Carly Brantmeyer, David Hyatt, Life Radio Ministries, CD Morris, Annette Mantooth, Art Spillers,Cherie Lucas, Deb Thompson, Hannah Mantooth, Kimberly Crow, Laura Kline, Melissa Mancuso, David McWhite, Michelle Discavage, Andrew Peterson, Kelly Waldrop, Carrie Carper, Jane Tuttle, Kelley Waller, Patrick McGill, RachelHenkel, William Lopez, Kim Pollard, Louella Venable, Poplar Springs Drive Baptist Church, Sheree Breech, Jenny Hopper, Aurora Gregory,

Kim Zajan, Rebecca Drieling, Stephanie Brady-Harding, Chris Loomis, Rosalie Brown, Michelle Huxford, Sarah Williams, Valerie Miller, Melinda Stembridge, Careless In the Care of God, Dana Nobles, Dustin Varn, Etta Marflak, Patti Lechner, Randy Langley, Anna Floit, April Best, Bonnie Horton, Heather Osterman, Sevana Petrosian, Jana Morris, Lisha Thigpen, Gloria Spillers, Stacy Taylor, Zoe Discavage, Ronald Cantrell, Randy Huffstetler, Don Durden, David Medley, Linda Myrick, James Brown, Melinda Williamson, Stacy Osterman, Susan Jennings, TabithaWestbrook, Tracy Stembridge, Balloon Creations, Brian Horton, Amy Howard, Jennifer Franco, Mike Robinson, Rebecca Ramsland, Lisa Harvey, David Loser, Dan Harding, Joel Martin, Tammy Helfrich, Virginia Blanton, Rosemarie Lones, Apryl Carrigan, Greta Burck, Christa Newell, Carol Corson, Cristy Davis, Feel Your Best, LLC (FYB), June Whitaker, Amanda Fischer, Stephanie Miranda, Natasha Bowles, Jonathan Cook, Stephanie Brown, Evan Chaisson

Contact Information

If you are not already a part of our family, I would encourage you to check out our website:

www.nolongerorphans.org

From the site you can make one time or recurring gifts. You can help fund our food, medical, or general budget. You can make a one time gift in honor of a loved one. You can help with some of our one time projects like building the school we mentioned.

On the site you can also check out our updates of our home and all the boys.

Feel free to also send us an email or a good old fashion snail mail latter. Our address are

info@nolongerorphans.org

No Longer Orphans
PO BOX 745185
Arvada, CO. 80005.

Made in the USA
Columbia, SC
02 October 2018